A Sound Engineer's Guide to Audio Test and Measurement

Glen Ballou

AMSTERDAM • BOSTON • HEIDELBERG • LONDON
NEW YORK • OXFORD • PARIS • SAN DIEGO
SAN FRANCISCO • SINGAPORE • SYDNEY • TOKYO

Academic Press is an Imprint of Elsevier

Focal Press is an imprint of Elsevier
30 Corporate Drive, Suite 400, Burlington, MA 01803, USA
Linacre House, Jordan Hill, Oxford OX2 8DP, UK

Library of Congress Cataloging-in-Publication Data
Application submitted

British Library Cataloguing-in-Publication Data
A catalogue record for this book is available from the British Library.

ISBN: 978-0-240-81265-6

For information on all Focal Press publications
visit our website at www.elsevierdirect.com

Typeset by diacriTech, India

09 10 11 5 4 3 2 1

Printed in the United States of America

Contents

Section 1
Test and Measurement

Pat Brown

Section 2
What's the Ear For? How to Protect It

Les Blomberg and Noland Lewis

Section 3
Fundamentals and Units of Measurement

Glen Ballou

Section 1

Test and Measurement

Pat Brown

1.1 TEST AND MEASUREMENT

Technological advancements in the last two decades have given us a variety of useful measurement tools, and most manufacturers of these instruments provide specialized training on their use. This chapter will examine some principles of test and measurement that are common to virtually all measurement systems. If the measurer understands the principles of measurement, then most any of the mainstream measurement tools will suffice for the collection and evaluation of data. The most important prerequisite to performing meaningful sound system measurements is that the measurer has a solid understanding of the basics of audio and acoustics. The question "How do I perform a measurement?" can be answered much more easily than "What should I measure?" This chapter will touch on both, but readers will find their measurement skills will relate directly to their understanding of the basic physics of sound and the factors that produce good sound quality. The whole of this book will provide much of the required information.

A Sound Engineer's Guide to Audio Test and Measurement

1.1.1 Why Test?

Sound systems must be tested to assure that all components are functioning properly. The test and measurement process can be subdivided into two major categories: electrical tests and acoustical tests. Electrical testing mainly involves voltage and impedance measurements made at component interfaces. Current can also be measured, but since the setup is inherently more complex it is usually calculated from knowledge of the voltage and impedance using Ohm's Law. Acoustical tests are more complex by nature, but share the same fundamentals as electrical tests in that some time varying quantity (usually pressure) is being measured. The main difference between electrical and acoustical testing is that the interpretation of the latter must deal with the complexities of 3D space, not just amplitude versus time at one point in a circuit. In this chapter we will define a loudspeaker system as a number of components intentionally combined to produce a system that may then be referred to as a loudspeaker. For example, a woofer, dome tweeter, and crossover network are individual components, but can be combined to form a loudspeaker system. Testing usually involves the measurement of systems, although a system might have to be dissected to fully characterize the response of each component.

1.2 ELECTRICAL TESTING

There are numerous electrical tests that can be performed on sound system components in the laboratory. The measurement system must have specifications that exceed the equipment being measured. Field testing need not be as comprehensive and the tests can be performed with less sophisticated instrumentation. The purpose for electrical field testing includes:

1. To determine if all system components are operating properly
2. To diagnose electrical problems in the system, which are usually manifested by some form of distortion
3. To establish a proper gain structure

Electrical measurements can aid greatly in establishing the proper gain structure of the sound system. Electrical test instruments that the author feels are essential to the audio technician include:

- ac voltmeter
- ac millivoltmeter
- Oscilloscope
- Impedance meter
- Signal generator
- Polarity test set

It is important to note that most audio products have on-board metering and/or indicators that may suffice for setting levels, making measurements with stand-alone meters unnecessary. Voltmeters and impedance meters are often only necessary for troubleshooting a nonworking system, or checking the accuracy and calibration of the on-board metering.

There are a number of currently available instruments designed specifically for audio professionals that perform all of the functions listed. These instruments need to have bandwidths that cover the audible spectrum. Many general purpose meters are designed primarily for ac power circuits and do not fit the wide bandwidth requirement.

More information on electrical testing is included in the chapter on gain structure. The remainder of this chapter will be devoted to the acoustical tests that are required to characterize loudspeakers and rooms.

1.3 ACOUSTICAL TESTING

The bulk of acoustical measurement and analysis today is being performed by instrumentation that includes or is controlled by a personal computer. Many excellent systems are available, and the would-be measurer should select the one that best fits their specific needs. As with loudspeakers, there is no clear-cut best choice or one-size-fits-all instrument. Fortunately an understanding of the principles of operating one analyzer can usually be applied to another after a short indoctrination period. Measurement systems are like rental cars: you know what features are there, you just need to find them. In this chapter I will attempt to provide a sufficient overview of the various approaches to allow the reader to investigate and select a tool to meet his or her measurement needs and budget. The acoustical field testing of sound reinforcement systems mainly involves measurements of the sound pressure fluctuations produced by one or more loudspeakers at various locations in the space. Microphone positions are selected based on the information that is needed. This could be the on-axis position of a loudspeaker for system alignment purposes, or a listener seat for measuring the clarity or intelligibility of the system. Measurements must be made to properly calibrate the system, which can include loudspeaker crossover settings, equalization, and the setting of signal delays. Acoustic waveforms are complex by nature, making them difficult to describe with one number readings for anything other than broadband level.

1.3.1 Sound Level Measurements

Sound level measurements are fundamental to all types of audio work. Unfortunately, the question: "How loud is it?" does not have a simple answer. Instruments can easily measure sound pressures, but there are many ways to describe the results in ways relevant to human perception. Sound pressures are usually measured at a discrete listener position. The sound pressure level may be displayed as is, integrated over a time interval, or frequency weighted by an appropriate filter. Fast meter response times produce information about peaks and transients in the program material, while slow response times yield data that correlates better with the perceived loudness and energy content of the sound.

A sound level meter consists of a pressure sensitive microphone, meter movement (or digital display), and some supporting circuitry, Fig. 1.1. It is used to observe the sound pressure on a moment-by-moment basis, with the pressure displayed as a level

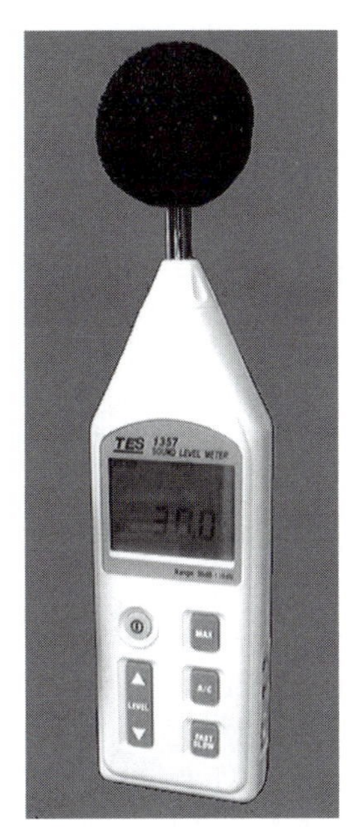

FIGURE 1.1 A sound level meter is basically a voltmeter that operates in the acoustic domain. Courtesy Galaxy Audio.

in decibels. Few sounds will measure the same from one instant to the next. Complex sounds such as speech and music will vary dramatically, making their level difficult to describe without a graph of level versus time, Fig. 1.2. A sound level meter is basically a voltmeter that operates in the acoustic domain.

Sound pressure measurements are converted into decibels ref. 0.00002 pascals. See Chapter 2: *Fundamentals of Audio and Acoustics,* for information about the decibel. Twenty micropascals are used as the reference because it is the threshold of pressure sensitivity for humans at midrange frequencies. Such measurements are referred to as *sound pressure level* or L_P (level of sound pressure) measurements, with L_P gaining acceptance among audio professionals because it is easily distinguished from L_W (sound power level) and L_I (sound intensity level) and a number of other L_X metrics used to describe sound levels. Sound pressure level is measured at a single point (the microphone position). Sound power measurements must consider all of the radiated sound from

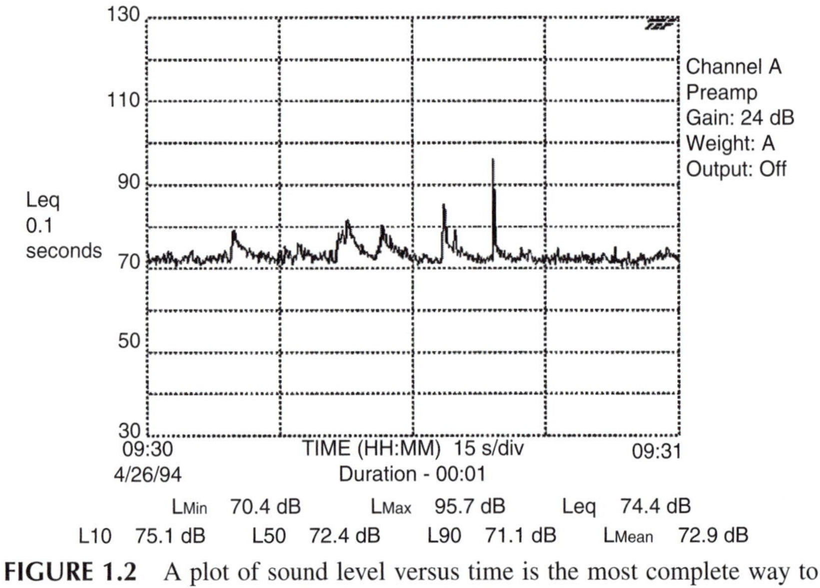

FIGURE 1.2 A plot of sound level versus time is the most complete way to record the level of an event. Courtesy Gold Line.

a device, and sound intensity measurements must consider the sound power flowing through an area. Sound power and sound intensity measurements are usually performed by acoustical laboratories rather than in the field, so neither is considered in this chapter. All measurements described in this chapter will be measurements of sound pressures expressed as levels in dB ref. 0.00002 Pa.

Sound level measurements must usually be processed for the data to correlate with human perception. Humans do not hear all frequencies with equal sensitivity, and to complicate things further our response is dependent on the level that we are hearing. The well-known Fletcher-Munson curves describe the frequency/level characteristics for an average listener (see Chapter 2). Sound level measurements are passed through weighting filters that make the meter "hear" with a response similar to a human. Each scale correlates with human hearing sensitivity at a different range of levels. For a sound level measurement to be meaningful, the weighting scale that was used must be indicated, in addition to the response time of the meter. Here are some examples of meaningful (if not universally accepted) expressions of sound level:

- The system produced an L_P = 100 dBA (slow response) at mix position.
- The peak sound level was L_A = 115 dB at my seat.
- The average sound pressure level was 100 dBC at 30 ft.
- The loudspeaker produced a continuous L_P of 100 dB at one meter (assumes no weighting used).
- The equivalent sound level L_{EQ} was 90 dBA at the farthest seat.

Level specifications should be stated clearly enough to allow someone to repeat the test from the description given. Because of the large differences between the weighting scales, it is meaningless to specify a sound level without indicating the scale that was used. An event that produces an L_P = 115 dB using a C scale may only measure as an L_P = 95 dB using the A scale.

The measurement distance should also be specified (but rarely is). Probably all sound reinforcement systems produce an L_P = 100 dB at some distance, but not all do so at the back row of the audience!

L_{PK} is the level of the highest instantaneous peak in the measured time interval. Peaks are of interest because our sound system components must be able to pass them without clipping them. A peak that is clipped produces high levels of harmonic distortion that degrade sound quality. Also, clipping reduces the crest factor of the waveform, causing more heat to be generated in the loudspeaker causing premature failure. Humans are not extremely sensitive to the loudness of peaks because our auditory system integrates energy over time with regard to loudness. We are, unfortunately, susceptible to damage from peaks, so they should not be ignored. Research suggests that it takes the brain about 35 ms to process sound information (frequency-dependent), which means that sound events closer together than this are blended together with regard to loudness. This is why your voice sounds louder in a small, hard room. It is also why the loudness of the vacuum cleaner varies from room to room. Short interval reflections are integrated with the direct sound by the ear/brain system. Most sound level meters have slow and fast settings that change the response time of the meter. The slow setting of most meters indicates the approximate root-mean-square sound level. This is the effective level of the signal, and should correlate well with its perceived loudness.

A survey of audio practitioners on the Syn-Aud-Con email discussion group revealed that most accept an $L_P = 95$ dBA (slow response) as the maximum acceptable sound level of a performance at any listener seat for a broad age group audience. The A-weighting is used because it considers the sound level in the portion of the spectrum where humans are most easily annoyed and damaged. The slow response time allows the measurement to ignore short duration peaks in the program. A measurement of this type will not indicate true levels for low-frequency information, but it is normally the mid-frequency levels that are of interest.

There exist a number of ways to quantify sound levels that are measured over time. They include:

- L_{PK}—the maximum instantaneous peak recorded during the span.
- L_{EQ}—the equivalent level (the integrated energy over a specified time interval).
- L_N—where L is the level exceeded N percent of the time.
- L_{DEN}—a special scale that weights the gathered sound levels based on the time of day. DEN stands for day-evening-night.
- **Dose**—a measure of the total sound exposure.

A variety of instruments are available to measure sound pressure levels, ranging from the simple sound level meter (SLM) to sophisticated data-logging equipment. SLMs are useful for making quick checks of sound levels. Most have at least an A- and C-weighting scale, and some have octave band filters that allow band-limited measurements. A useful feature on an SLM is an output jack that allows access to the measured data in the form of an ac voltage. Software applications are available that can log the meter's response versus time and display the results in various ways. A plot of sound level versus time is the most complete way to record the level of an event. Fig. 1.2 is such a measurement. Note that a start time and stop time are specified. Such measurements usually provide statistical summaries for the recorded data. An increasing number of venues monitor the levels of performing acts in this manner due to growing concerns about litigation over hearing damage to patrons. SLMs vary dramatically in price, depending on quality and accuracy.

All sound level meters provide accurate indications for relative levels. For absolute level measurements a calibrator must be used to calibrate the measurement system. Many PC-based measurement systems have routines that automate the calibration process. The calibrator is placed on the microphone, Fig. 1.3, and the calibrator level (usually 94 or 114 dB ref. 20 μPa) is entered into a data field. The measurement tool now has a true level to use as a reference for displaying measured data.

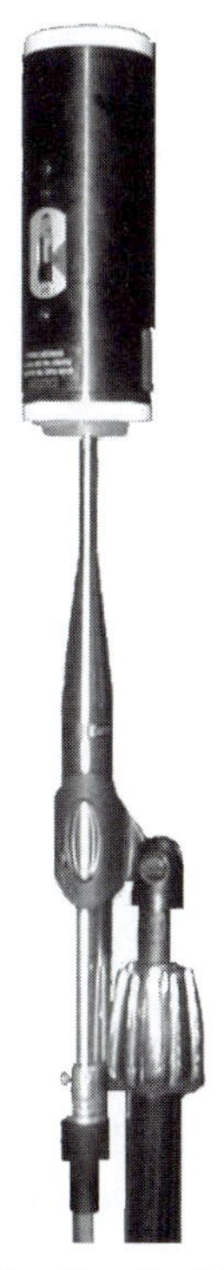

FIGURE 1.3 A calibrator must be fitted with a disc to provide a snug fit to the microphone. Most microphone manufacturers can provide the disc.

Noise criteria ratings provide a one-number specification for allowable levels of ambient noise. Sound level measurements are performed in octave bands, and the results are plotted on the chart shown in Fig. 1.4. The NC rating is read on the right vertical axis. Note that the NC curve is frequency-weighted. It permits an increased level of low-frequency noise, but becomes more stringent at higher frequencies. A sound system specification should include an NC rating for the space, since excessive ambient noise will reduce system clarity and require additional acoustic gain. This must be considered when designing the sound system. Instrumentation is available to automate noise criteria measurements.

1.3.1.1 Conclusion

Stated sound level measurements are often so ambiguous as to become meaningless. When stating a sound level, it is important to indicate:

1. The sound pressure level
2. Any weighting scale used
3. Meter response time (fast, slow or other)

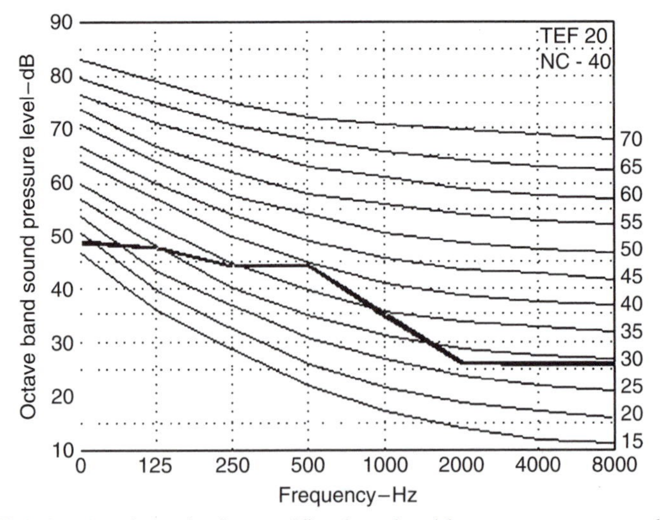

FIGURE 1.4 A noise criteria specification should accompany a sound system specification.

4. The distance or location at which the measurement was made
5. The type of program measured (i.e., music, speech, ambient noise)

Some correct examples:

- "The house system produced 90 dBA-Slow in section C for broadband program."
- "The monitor system produced 105 dBA-Slow at the performer's head position for broadband program."
- "The ambient noise with room empty was NC-35 with HVAC running."

In short, if you read the number and have to request clarification then sufficient information has not been given. As you can see, one-number SPL ratings are rarely useful.

All sound technicians should own a sound level meter, and many can justify investment in more elaborate systems that provide statistics on the measured sound levels. From a practical perspective, it is a worthwhile endeavor to train one's self to recognize various sound levels without a meter, if for no other reason than to find an exit in a venue where excessive levels exist.

1.3.2 Detailed Sound Measurements

The response of a loudspeaker or room must be measured with appropriate frequency resolution to be characterized. It is also important for the measurer to understand what the appropriate response should be. If the same criteria were applied to a loudspeaker as to an electronic component such as a mixer, the optimum response would be a flat (minimal variation) magnitude and phase response at all frequencies within the required pass band of the system. In reality, we are usually testing loudspeakers to make sure that they are operating at their fullest potential. While flat magnitude and phase response are a noble objective, the physical reality is that we must often settle for far less in terms of accuracy. Notwithstanding, even with their inherent inaccuracies, many loudspeakers do an outstanding job of delivering speech or music to the audience. Part of the role of the measurer is to determine if the response of the loudspeaker or room is inhibiting the required system performance.

1.3.2.1 Sound Persistence in Enclosed Spaces

Sound system performance is greatly affected by the sound energy persistence in the listening space. One metric that is useful for describing this characteristic is the reverberation time, T_{30}. The T_{30} is the time required for an interrupted steady-state sound source to decay to inaudibility. This will be about 60 dB of decay in most auditoriums with controlled ambient noise floors. The T_{30} designation comes from the practice of measuring 30 dB of decay and then doubling the time interval to get the time required for 60 dB of decay. A number of methods exist for determining the T_{30}, ranging from simple listening tests to sophisticated analytical methods. Fig. 1.5 shows a simple gated-noise test that can provide sufficient accuracy for designing systems. The bursts for this test can be generated with a WAV editor. Bursts of up to 5 seconds for each of 8 octave bands should be generated. Octave band-limited noise is played into the space through a low directivity loudspeaker. The noise is gated on for 1 second and off for 1 second. The room decay is evaluated during the off span. If it decays completely before the next burst, the T_{30} is less than 1 second. If not, the next burst should be on for 2 seconds and off for 2 seconds. The measurer simply keeps advancing to the next track until the room completely decays in the off span, Figs. 1.6, 1.7, and 1.8. The advantages of this method include:

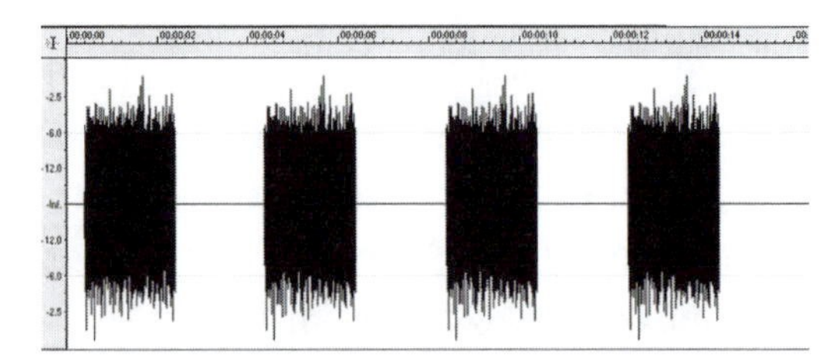

FIGURE 1.5 Level versus time plot of a 1 octave band gated burst (2 second duration).

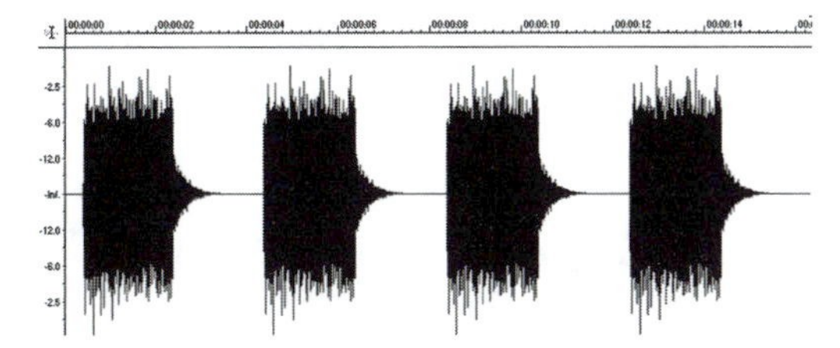

FIGURE 1.6 A room with $RT_{60} < 2$ seconds.

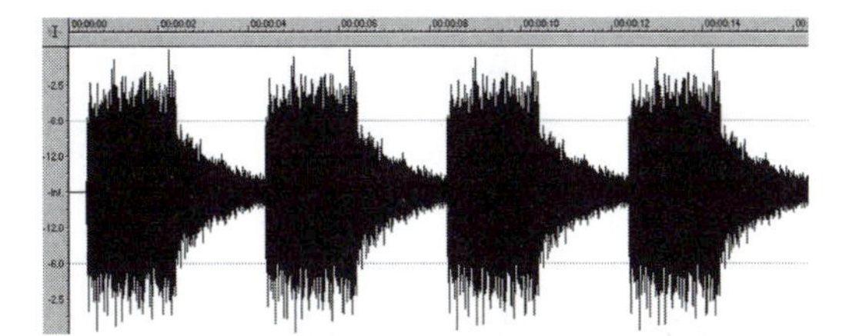

FIGURE 1.7 A room with $RT_{60} > 2$ seconds.

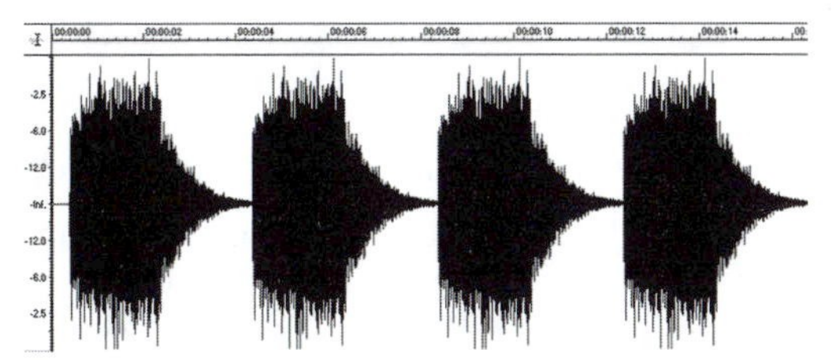

FIGURE 1.8 A room with $RT_{60} = 2$ seconds.

1. No sophisticated instrumentation is required.
2. The measurer is free to wander the space.
3. The nature of the decaying field can be judged.
4. A group can perform the measurement.

A test of this type is useful as a prelude to more sophisticated techniques.

1.3.2.2 Amplitude versus Time

Fig. 1.9 shows an audio waveform displayed as amplitude versus time. This representation is especially meaningful to humans since it can represent the motion of the eardrum about its resting position. The waveform shown is of a male talker recorded in an anechoic

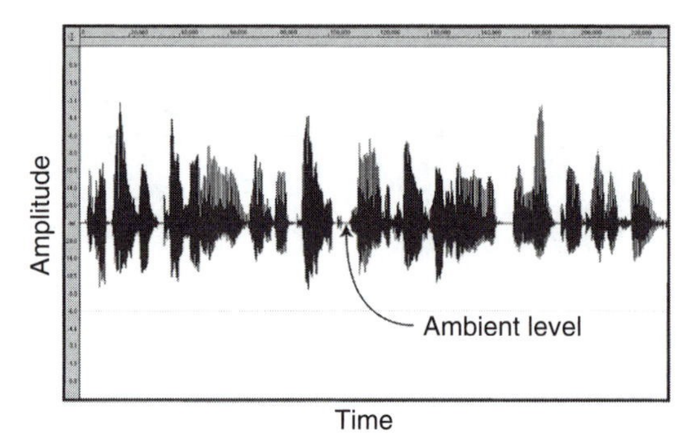

FIGURE 1.9 Amplitude versus time plot of a male talker made in an anechoic environment.

(echo-free) environment. The 0 line represents the ambient (no signal) state of the medium being modulated. This would be ambient atmospheric pressure for an acoustical wave, or zero volts or a dc offset for an electrical waveform measured at the output of a system component.

Fig. 1.10 shows the same waveform, but this time played over a loudspeaker into a room and recorded. The waveform has now been encoded (convolved) with the response of the loudspeaker and room. It will sound completely different than the anechoic version.

Fig. 1.11 shows an impulse response and Fig. 1.12 shows the envelope-time curve (ETC) of the loudspeaker and room. It is essentially the difference between Fig. 1.9 and Fig. 1.10 that fully characterizes any effect that the loudspeaker or room has on the electrical signal fed to the loudspeaker and measured at that point in space. Most measurement systems attempt to measure the impulse response, since knowledge of the impulse response of a system allows its effect on any signal passing through it to be determined, assuming the system is linear and time invariant. This effect is called the transfer function of the system and includes both magnitude (level) and phase (timing) information for each frequency in the pass band. Both the loudspeaker and room can be considered filters that the energy must pass through en route to the listener. Treating them as filters allows their responses to be measured

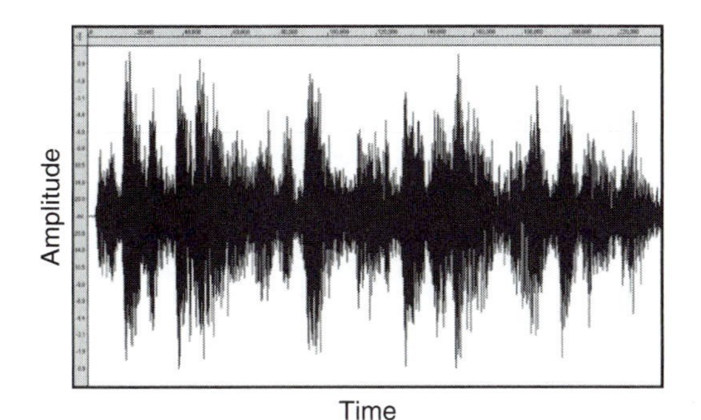

FIGURE 1.10 The voice waveform after encoding with the room response.

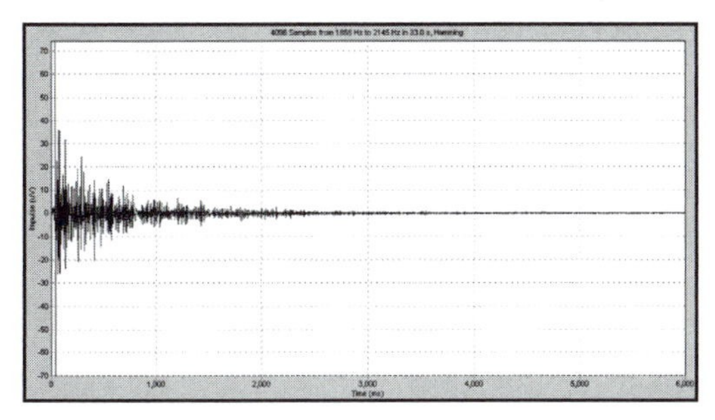

FIGURE 1.11 The impulse response of the acoustic environment.

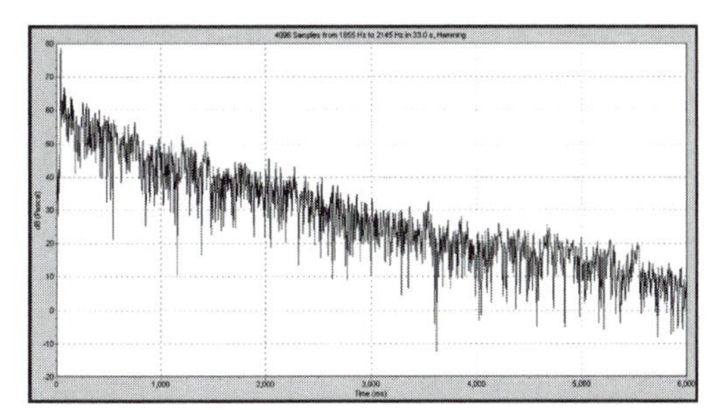

FIGURE 1.12 The envelope-time curve (ETC) of the same environment. It can be derived from the impulse response.

and displayed, and provides an objective benchmark to evaluate their effect. It also opens loudspeakers and rooms to evaluation by electrical network analysis methods, which are generally more widely known and better developed than acoustical measurement methods.

1.3.2.3 The Transfer Function

The effect that a filter has on a waveform is called its *transfer function.* A transfer function can be found by comparing the input signal and output signal of the filter. It matters little if the filter is an electronic component, loudspeaker, room, or listener. The time domain behavior of a system (impulse response) can be displayed in the frequency domain as a spectrum and phase (transfer function). Either the time or frequency description fully describes the filter. Knowledge of one allows the determination of the other. The mathematical map between the two representations is called a transform. Transforms can be performed at amazingly fast speeds by computers. Fig. 1.13 shows a domain chart that provides a map between various representations of a system's response. The measurer must remember that the responses being measured and displayed on the analyzer are dependent on the test

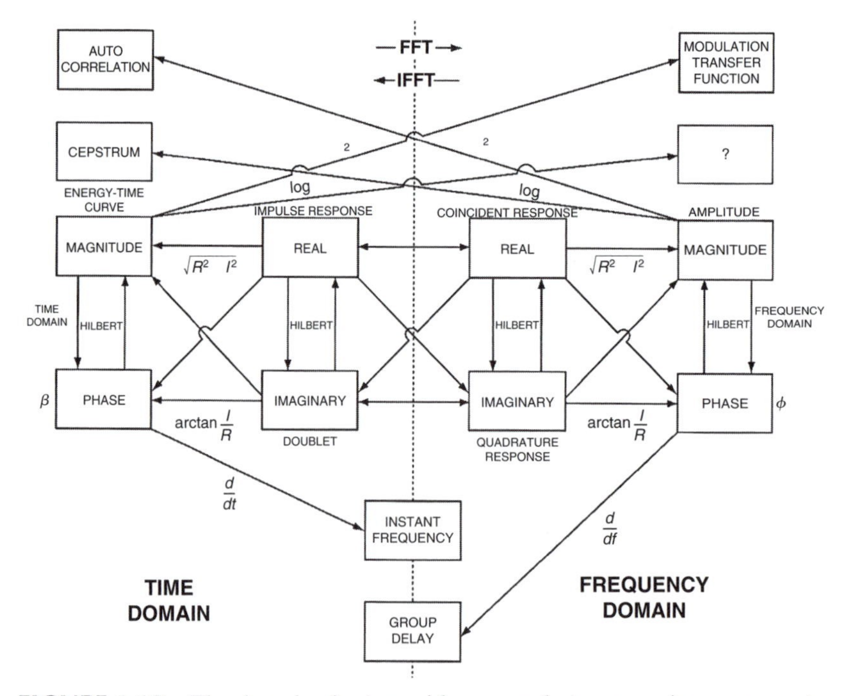

FIGURE 1.13 The domain chart provides a map between various representations of a system response. Courtesy Brüel and Kjaer.

stimulus used to acquire the response. Appropriate stimuli must have adequate energy content over the pass band of the system being measured. In other words, we can't measure a subwoofer using a flute solo as a stimulus. With that criteria met, the response measured and displayed on the analyzer is independent of the program material that passes through a linear system. Pink noise and sine sweeps are common stimuli due to their broadband spectral content. In other words, the response of the system doesn't change relative to the nature of the program material. For a linear system, the transfer function is a summary that says, "If you put energy into this system, this is what will happen to it."

The domain chart provides a map between various methods of displaying the system's response. The utility of this is that it allows measurement in either the time or frequency domain. The alternate view can be determined mathematically by use of a transform. This allows frequency information to be determined with a time domain measurement, and time information to be determined by a frequency domain measurement. This important inverse relationship between time and frequency can be exploited to yield many possible ways of measuring a system and/or displaying its response. For instance, a noise immunity characteristic not attainable in the time domain may be attainable in the frequency domain. This information can then be viewed in the time domain by use of a transform. The Fourier Transform and its inverse are commonly employed for this purpose. Measurement programs like Arta can display the signal in either domain, Fig. 1.14.

1.3.3 Measurement Systems

Any useful measurement system must be able to extract the system response in the presence of noise. In some applications, the signal-to-noise requirements might actually determine the type of analysis that will be used. Some of the simplest and most convenient tests have poor signal-to-noise performance, while some of the most complex and computationally demanding methods can measure under almost any conditions. The measurer must choose

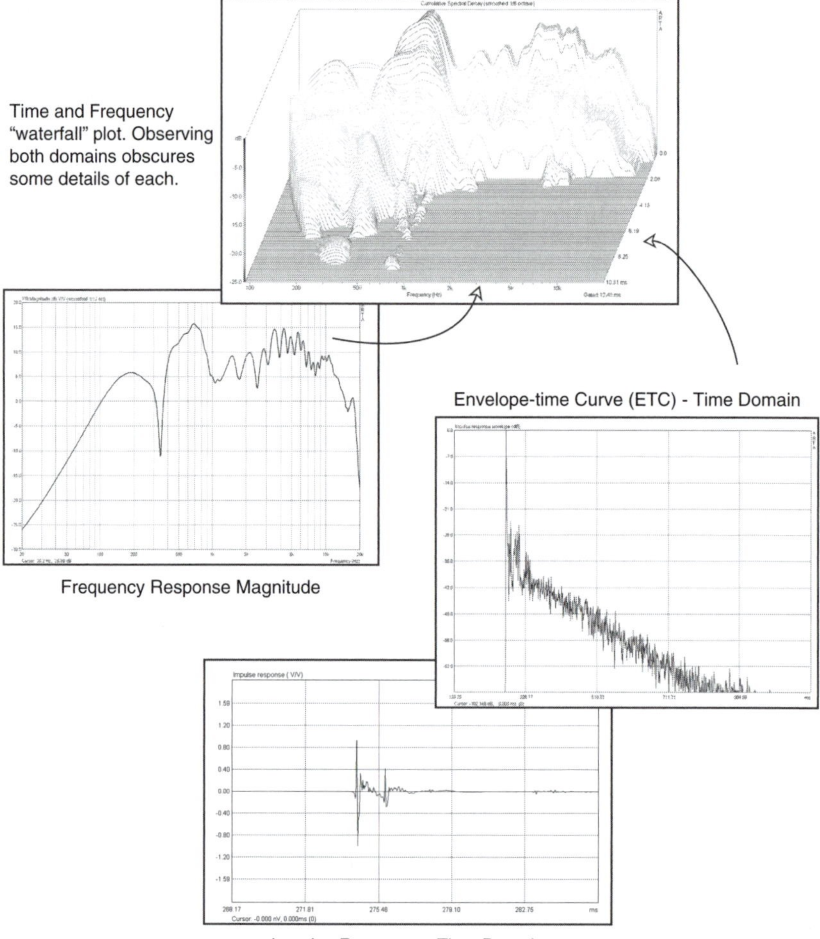

FIGURE 1.14 The FFT can be used to view the spectral content of a time domain measurement, Arta 1.2.

the type of analysis with these factors in mind. It is possible to acquire the impulse response of a filter without using an impulse. This is accomplished by feeding a known broadband stimulus into the filter and reacquiring it at the output. A complex comparison of the two signals (mathematical division) yields the transfer function, which is displayed in the frequency domain as a magnitude and phase, or inverse-transformed for display in the time domain as an impulse response. The impulse response of a system answers the question, "If I feed a perfect impulse into this system, when will the energy exit the system?" A knowledge of "when" can characterize a system. After transformation, the spectrum or frequency response is displayed on a decibel scale. A phase plot shows the phase response of the device under test, and any phase shift versus frequency becomes apparent. If an impulse response is a measure of when, we might describe a frequency response as a measure of what. In other words, "If I input a broadband stimulus (all frequencies) into the system, what frequencies will be present at the output of the system and what will their phase relationship be?" A transfer function includes both magnitude and phase information.

1.3.3.1 Alternate Perspectives

The time and frequency views of a system's response are mutually exclusive. By definition the time period of a periodic event is

$$T = \frac{1}{f} \tag{1.1}$$

where:

T is time in seconds,
f is frequency in hertz.

Since time and frequency are reciprocals, a view of one excludes a view of the other. Frequency information cannot be observed on an impulse response plot, and time information cannot be observed on a magnitude/phase plot. Any attempt to view both simultaneously will obscure some of the detail of both. Modern analyzers allow the measurer to switch between the time and frequency perspectives to extract information from the data.

1.3.4 Testing Methods

Compared to other components in the sound system, the basic design of loudspeakers and compression drivers has changed relatively little in the last 50 years. At over a half-century since their invention, we are still pushing air with pistons driven by voice coils suspended in magnetic fields. But the methods for measuring their performance have improved steadily since computers can now efficiently perform digital sampling and signal processing, and execute transforms in fractions of a second. Extremely capable measurement systems are now accessible and affordable to even the smallest manufacturers and individual audio practitioners. A common attribute of systems suitable for loudspeaker testing is the ability to make reflection-free measurements indoors, without the need for an anechoic chamber. Anechoic measurements in live spaces can be accomplished by the use of a time window that allows the analyzer to collect the direct field response of the loudspeaker while ignoring room reflections. Conceptually, a time window can be thought of as an accurate switch that can be closed as the desired waves pass the microphone and opened prior to the arrival of undesirable reflections from the environment. A number of implementations exist, each with its own set of advantages and drawbacks. The potential buyer must understand the trade-offs and choose a system that offers the best set of compromises for the intended application. Parameters of interest include signal-to-noise ratios, speed, resolution, and price.

1.3.4.1 FFT Measurements

The Fourier Transform is a mathematical filtering process that determines the spectral content of a time domain signal. The Fast Fourier Transform, or FFT, is a computationally efficient version of the same. Most modern measurement systems make use of the computer's ability to quickly perform the FFT on sampled data. The cousin to the FFT is the IFFT, or Inverse Fast Fourier Transform. As one might guess, the IFFT takes a frequency domain signal as its input and produces a time domain signal. The FFT and IFFT form the bedrock of modern measurement systems. Many fields outside of audio use the FFT to analyze time records for periodic activity, such as utility companies to find peak usage times or an investment firm to investigate cyclic stock market behavior. Analyzers that use the Fast Fourier Transform to determine the spectral content of a time-varying signal are collectively called FFTs. If a broadband stimulus is used, the FFT can show the spectral response of the device under test (DUT). One such stimulus is the unit impulse, a signal of theoretically infinite amplitude and infinitely small time duration. The FFT of such a stimulus is a straight, horizontal line in the frequency domain.

The time-honored hand clap test of a room is a crude but useful form of impulse response. The hand clap is useful for casual observations, but more accurate and repeatable methods are usually required for serious audio work. The drawbacks of using impulsive stimuli to measure a sound system include:

1. Impulses can drive loudspeakers into nonlinear behavior.
2. Impulse responses have poor signal-to-noise ratios, since all of the energy enters the system at one time and is reacquired over a longer span of time along with the noise from the environment.
3. There is no way to create a perfect impulse, so there will always be some uncertainty as to whether the response characteristic is that of the system, the impulse, or some nonlinearity arising from impulsing a loudspeaker.

Even with its drawbacks, impulse testing can provide useful information about the response of a loudspeaker or room.

1.3.4.2 Dual-Channel FFT

When used for acoustic measurements, dual-channel FFT analyzers digitally sample the signal fed to the loudspeaker, and also digitally sample the acoustic signal from the loudspeaker at the output of a test microphone. The signals are then compared by division, yielding the transfer function of the loudspeaker. Dual-channel FFTs have the advantage of being able to use any broadband stimulus as a test signal. This advantage is offset somewhat by poorer signal-to-noise performance and stability than other types of measurement systems, but the performance is often adequate for many measurement chores. Pink noise and swept sines provide much better stability and noise immunity. It is a computationally intense method since both the input and output signal must be measured simultaneously and compared, often in real time. For a proper comparison to yield a loudspeaker transfer function, it is important that the signals being compared have the same level, and that any time offsets between the two signals be removed. Dual-channel FFT analyzers have set up routines that simplify the establishment of these conditions. Portable computers have A/D converters as part of their on-board sound system, as well as a microprocessor to perform the FFT. With the appropriate software and sound system interface they form a powerful, low-cost and portable measurement platform.

1.3.4.3 Maximum-Length Sequence

The maximum-length sequence (MLS) is a pseudorandom noise test stimulus. The MLS overcomes some of the shortcomings of the dual-channel FFT, since it does not require that the input signal to the system be measured. A binary string (ones and zeros) is fed to the device under test while simultaneously being stored for future correlation with the loudspeaker response acquired by the test microphone. The pseudorandom sequence has a white spectrum (equal energy per Hz), and is exactly known and exactly repeatable. Comparing the input string with the string acquired by the test microphone yields the transfer function of the system. The advantage of the MLS is its excellent noise immunity and fast measurement time, making it a favorite of loudspeaker designers. A disadvantage is that the noiselike stimulus can be annoying, sometimes requiring that measurements be done after hours. The use of MLS has waned in recent years to log-swept sine measurements made on dual-channel FFT analyzers.

1.3.4.4 Time-Delay Spectrometry (TDS)

TDS is a fundamentally different method of measuring the transfer function of a system. Richard Heyser, a staff scientist at the Jet Propulsion Laboratories, invented the method. An anthology of Mr. Heyser's papers on TDS is available in the reference. Both the dual-channel FFT and MLS methods involve digital sampling of a broadband stimulus. TDS uses a method borrowed from the world of sonar, where a single-frequency sinusoidal "chirp" signal is fed to the system under test. The chirp slowly sweeps through the frequencies being measured, and is reacquired with a tracking filter by the TDS analyzer. The reacquired signal is then mixed with the outgoing signal, producing a series of sum and difference frequencies, each frequency corresponding to a different arrival time of sound at the microphone. The difference frequencies are transformed to the time domain with the appropriate transform, yielding the envelope-time curve (ETC) of the system under test. TDS is based on the frequency domain, allowing the tracking filter to be tuned to the desired signal while ignoring signals outside of its bandwidth. TDS offers excellent noise immunity, allowing good data to be collected under near-impossible measurement conditions. Its downside is that good low-frequency resolution can be difficult to obtain without extended measurement times, plus the correct selection of measurement parameters requires a knowledgeable user. In spite of this, it is a favorite among contractors and consultants, who must often perform sound system calibrations in the real world of air conditioners, vacuum cleaners, and building occupants.

While other measurement methods exist, the ones outlined above make up the majority of methods used for field and lab testing of loudspeakers and rooms. Used properly, any of the methods can provide accurate and repeatable measured data. Many audio professionals have several measurement platforms and exploit the strong points of each when measuring a sound system.

1.3.5 Preparation

There are many measurements that can be performed on a sound system. A prerequisite to any measurement is to answer the following questions:

1. What am I trying to measure?
2. Why am I trying to measure it?

3. Is it audible?
4. Is it relevant?

Failure to consider these questions can lead to hours of wasted time and a hard drive full of meaningless data. Even with the incredible technologies that we have available to us, the first part of any measurement session is to listen. It can take many hours to determine what needs to be measured to solve a sound system problem, yet the actual measurement itself can often be completed in seconds. Using an analogy from the medical field, the physician must query the patient at length to discover the ailment. The more that is known about the ailment, the more specific and relevant the tests that can be run for diagnosis. There is no need to test for tonsillitis if the problem is a sore back!

1. What am I measuring? A fundamental decision that precedes a meaningful measurement is how much of the room's response to include in the measured data. Modern measurement systems have the ability to perform semianechoic measurements, and the measurer must decide if the loudspeaker, the room, or the combination needs to be measured. If one is diagnosing loudspeaker ailments, there is little reason to select a time window long enough to include the effects of late reflections and reverberation. A properly selected time window can isolate the direct field of the loudspeaker and allow its response to be evaluated independently of the room. If one is trying to measure the total decay time of the room, the direct sound field becomes less important, and a microphone placement and time window are selected to capture the entire energy decay. Most modern measurement systems acquire the complete impulse response, including the room decay, so the choice of the time window size can be made after the fact during post processing.
2. Why am I measuring? There are several reasons for performing acoustic measurements in a space. An important reason for the system designer is to characterize the listening environment.

Is it dead? Is it live? Is it reverberant? These questions must be considered prior to the design of a sound system for the space. While the human hearing system can provide the answers to these questions, it cannot document them and it is easily deceived. Measurements might also be performed to document the performance of an existing system prior to performing changes or adding room treatment. Customers sometimes forget how bad it once sounded after a new or upgraded system is in place for a few weeks.

The most common reason for performing measurements on a system is for calibration purposes. This can include equalization, signal alignment, crossover selection, and a multiplicity of other reasons. Since loudspeakers interact in a complex way with their environment, the final phase of any system installation is to verify system performance by measurement.

3. Is it audible? Can I hear what I am trying to measure? If one cannot hear an anomaly, there is little reason to attempt to measure it. The human hearing system is perhaps the best tool available for determining what should be measured about a sound system. The human hearing system can tell us that something doesn't sound right, but the cause of the problem can be revealed by measurement. Anything you can hear can be measured, and once it is measured it can be quantified and manipulated.
4. Is it relevant? Am I measuring something that is worth measuring? If one is working for a client, time is money. Measurements must be prioritized to focus on audible problems. Endless hours can be spent "chasing rabbits" by measuring details that are of no importance to the client. This is not necessarily a fruitless process, but it is one that should be done on your own time. I have on several occasions spent time measuring and documenting anomalies that had nothing to do with the customer's reason for calling me. All venues have problems that the owner is unaware of. Communication with the client is the best way to avoid this pitfall.

1.3.5.1 Dissecting the Impulse Response

The audio practitioner is often faced with the dilemma of determining whether the reason for bad sound is the loudspeaker system, the room, or an interaction of the two. The impulse response can hold the answer to these and other perplexing questions. The impulse response in its amplitude versus time display is not particularly useful for other than determining the polarity of a system component, Fig. 1.15. A better representation comes from squaring impulse response (making all deflections positive) and displaying the square root of the result on a logarithmic vertical scale. This log-squared response allows the relative levels of energy arrivals to be compared, Fig. 1.16.

1.3.5.2 The Envelope-Time Curve

Another useful way of viewing the impulse response is in the form of the envelope-time curve, or ETC. The ETC is also a contribution of Richard Heyser.[2] It takes the real part of the impulse response and combines it with a 90 degrees phase shifted version of the same, Fig. 1.17. One way to get the shifted version is to use the Hilbert Transform. The complex combination of these two signals yields a time domain waveform that is often easier to interpret than the impulse response. The ETC can be loosely thought of as a smoothing function for the log-squared response, showing the envelope of the data. This can be more revealing as to the audibility of an

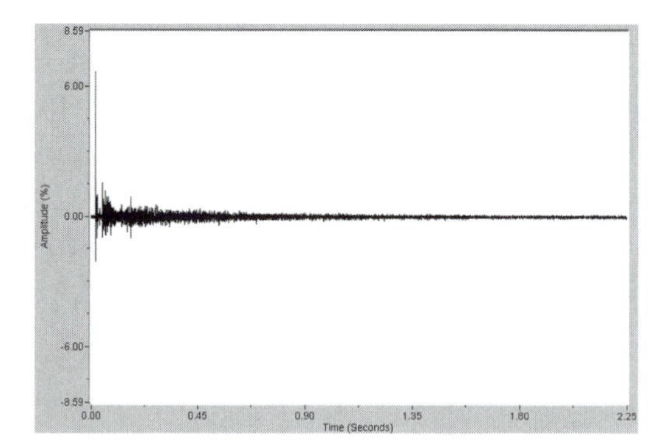

FIGURE 1.15 The impulse response, SIA-SMAART.

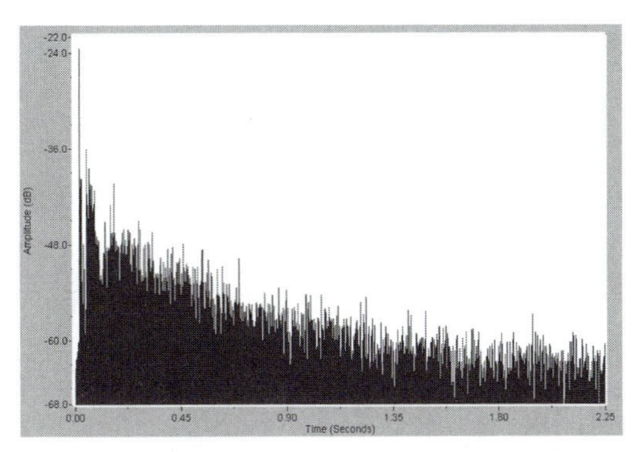

FIGURE 1.16 The log-squared response, SIA-SMAART.

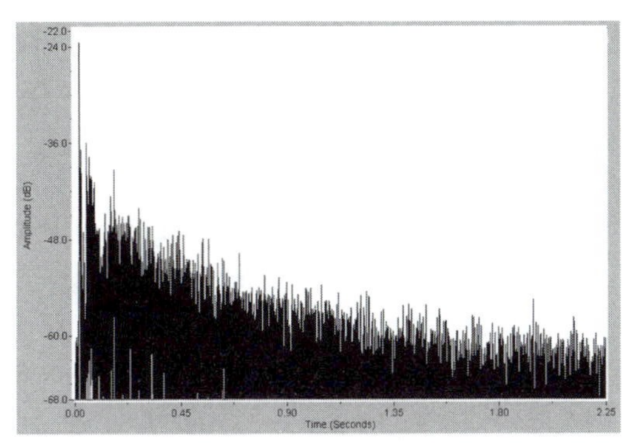

FIGURE 1.17 The envelope-time curve (ETC), SIA-SMAART.

event. The impulse response, log-squared response, and energy-time curve are all different ways to view the time domain data.

1.3.5.3 A Global Look

When starting a measurement session, a practical approach is to first take a global look and measure the complete decay of the room. The measurer can then choose to ignore part of the time record by using a time window to isolate the desired part during post processing. The length of the time window can be increased to include the effects of more of the energy returned by the room. The time window can also be used to isolate a reflection and view its spectral content. Just like your life span represents a time window in human history, a time window can be used to isolate parts of the impulse response.

1.3.5.4 Time Window Length

The time domain response can be divided to identify the portion that can be attributed to the loudspeaker and that which can be attributed to the room. It must be emphasized that there is a rather gray and frequency-dependent line between the two, but for this discussion we will assume that we can clearly separate them. The direct field is the energy that arrives at the listener prior to any reflections from the room. The division is fairly distinct if neither the loudspeaker nor microphone is placed near any reflecting surfaces, which, by the way, is a good system design practice. At long wavelengths (low frequencies) the direct field may include the effects of boundaries near the loudspeaker and microphone. As frequency increases, the sound from the loudspeaker becomes less affected by boundary effects (due in part to increased directivity) and can be measured independently of them. Proper loudspeaker placement produces a time gap between the sound energy arrivals from the loudspeaker and the later arriving room response. We can use this time gap to aid in selecting a time window to separate the loudspeaker response from the room response and diagnosing system problems.

1.3.5.5 Acoustic Wavelengths

Sound travels in waves. The sound waves that we are interested in characterizing have a physical size. There will be a minimum time span required to observe the spectral response of a waveform. The minimum required length of time to view an acoustical event is determined by the longest wavelength (lowest frequency) present in the event. At the upper limits of human hearing, the wavelengths are only a few millimeters in length, but as frequency decreases the waves become increasingly large. At the lowest frequencies that humans hear, the wavelengths are many meters long, and can actually be larger than the listening (or measurement) space. This makes it difficult to measure low frequencies from a loudspeaker independently of the listening space, since low frequencies radiated from a loudspeaker interact (couple) with the surfaces around them. In an ideally positioned loudspeaker, the first energy arrival from the loudspeaker at mid- and high frequencies has already dissipated prior to the arrival of reflections and can therefore often be measured independently of them. The human hearing system tends to fuse the direct sound from the loudspeaker with the early reflections from nearby surfaces with regard to level (loudness) and frequency (tone). It is usually useful to consider them as separate events, especially since the time offset between the direct sound and first reflections will be unique for each listening position. This precludes any type of frequency domain correction (i.e., equalization) of the room/loudspeaker response other than at frequencies where coupling occurs due to close proximity to nearby surfaces. While it is possible to compensate to some extent for room reflections at a point in space (acoustic echo cancellers used for conference systems), this correction cannot be extended to include an area. This inability to compensate for the reflected energy at mid/high frequencies suggests that their effects be removed from the loudspeaker's direct field response prior to meaningful equalization work by use of an appropriate time window.

1.3.5.6 Microphone Placement

A microphone is needed to acquire the sound radiated into the space from the loudspeaker at a discrete position. Proper microphone placement is determined by the type of test being performed. If one were interested in measuring the decay time of the room, it is usually best to place the microphone well beyond critical distance. This allows the build-up of the reverberant field to be observed as well as providing good resolution of the decaying tail. Critical distance is the distance from the loudspeaker at which the direct field level and reverberant field level are equal. It is described further in Section 1.3.5.7. If it's the loudspeaker's response that needs to be measured, then a microphone placement inside of critical distance will provide better data on some types of analyzers, since the direct sound field is stronger relative to the later energy returning from the room. If the microphone is placed too close to the loudspeaker, the measured sound levels will be accurate for that position, but may not accurately extrapolate to greater distances with the inverse-square law. As the sound travels farther, the response at a remote listening position may bear little resemblance to the response at the near field microphone position. For this reason, it is usually desirable to place the microphone in the far free field of the loudspeaker—not too close and not too far away. The approximate extent of the near field can be determined by considering that the path length difference from the measurement position (assumed axial) and the edge of the sound radiator should be less than ¼ wavelength at the frequency of interest. This condition is easily met for a small loudspeaker that is radiating low frequencies. Such devices closely approximate an ideal point source. As the frequency increases the condition becomes more difficult to satisfy, especially if the size of the radiator also increases. Large radiators (or groups of radiators) emitting high frequencies can extend the near field to very long distances. Line arrays make use of this principle to overcome the inverse-square law. In practice, small bookshelf loudspeakers can be accurately measured at a few meters. About 10 m is a common measurement distance for moderate-sized, full-range loudspeakers in a large space. Even greater distances are required for large devices radiating high frequencies. A general guideline is to not put the mic closer than three times the loudspeaker's longest dimension.

1.3.5.7 Estimate the Critical Distance D_C

Critical distance is easy to estimate. A quick method with adequate accuracy requires a sound level meter and noise source. Ideally, the noise source should be band limited, as critical distance is frequency-dependent. The 2 kHz octave band is a good place to start when measuring critical distance. Proceed as follows:

1. Energize the room with pink noise in the desired octave band from the sound source being measured. The level should be

at least 25 dB higher than the background noise in the same octave band.

2. Using the sound level meter, take a reading near the loudspeaker (at about 1 m) and on-axis. At this distance, the direct sound field will dominate the measurement.
3. Move away from the loudspeaker while observing the sound level meter. The sound level will fall off as you move farther away. If you are in a room with a reverberant sound field, at some distance the meter reading will quit dropping. You have now moved beyond critical distance. Measurements of the direct field beyond this point will be a challenge for some types of analysis. Move back toward the loudspeaker until the meter begins to rise again. You are now entering a good region to perform acoustic measurements on loudspeakers in this environment. The above process provides an estimate that is adequate for positioning a measurement microphone for loudspeaker testing. With a mic placement inside of critical distance, the direct field is a more dominant feature on the impulse response and a time window will be more effective in removing room reflections.

At this point it is interesting to wander around the room with the sound level meter and evaluate the uniformity of the reverberant field. Rooms that are reverberant by the classical definition will vary little in sound level beyond critical distance when energized with a continuous noise spectrum. Such spaces have low internal sound absorption relative to their volume.

1.3.5.8 Common Factors to All Measurement Systems

Let's assume that we wish to measure the impulse response of a loudspeaker/room combination. While it would not be practical to measure the response at every seat, it is good measurement practice to measure at as many seats as are required to prove the performance of the system. Once the impulse response is properly acquired, any number of post processes can be performed on the data to extract information from it. Most modern measurement systems make use of digital sampling in acquiring the response of the system. The fundamentals and prerequisites are not unlike the techniques used to make any digital recording, where one must be concerned with the level of an event and its time length. Some setup is required and some fundamentals are as follows:

1. The sampling rate must be fast enough to capture the highest frequency component of interest. This requires at least two samples of the highest frequency component. If one wished to measure to 20 kHz, the required sample rate would need to be at least 40 kHz. Most measurement systems sample at 44.1 kHz or 48 kHz, more than sufficient for acoustic measurements.
2. The time length of the measurement must be long enough to allow the decaying energy curve to flatten out into the room noise floor. Care must be taken to not cut off the decaying energy, as this will result in artifacts in the data, like a scratch on a phonograph record. If the sampling rate is 44.1 kHz, then 44,100 samples must be collected for each second of room decay. A 3-second room would therefore require 44.1 × 1000 × 3 or 128,000 samples. A hand clap test is a good way to estimate the decay time of the room and therefore the required number of samples to fully capture it. The time span of the measurement also determines the lowest frequency that can be resolved from the measured data, which is approximately the inverse of the measurement length. The sampling rate can be reduced to increase the sampling time to yield better low-frequency information. The trade-off is a reduction in the highest frequency that can be measured, since the condition outlined in step one may have been violated.
3. The measurement must have a sufficient signal-to-noise ratio to allow the decaying tail to be fully observed. This often requires that the measurement be repeated a number of times and the results averaged. Using a dual-channel FFT or MLS, the improvement in SNR will be 3 dB for each doubling of the number of averages. Ten averages is a good place to start, and this number can be increased or decreased depending on the

environment. The level of the test stimulus is also important. Higher levels produce improved SNR, but can also stress the loudspeaker.

4. Perform the test and observe the data. It should fill the screen from top left to bottom right and be fully decayed prior to reaching the right side of the screen. It should also be repeatable. Run the test several times to check for consistency. Background noise can dramatically affect the repeatability of the measurement and the validity of the data.

Once the impulse response is acquired, it can be further analyzed for spectral content, intelligibility information, decay time, etc. These are referred to as metrics, and some require some knowledge on the part of the measurer in properly placing markers (called *cursors*) to identify the parameters required to perform the calculations. Let us look at how the response of the loudspeaker might be extracted from the data just gathered.

The time domain data displays what would have resulted if an impulse were fed through the system. Don't try to correlate what you see on the analyzer with what you heard during the test. Most measurement systems display an impulse response that is calculated from a knowledge of the input and output signal to the system, and there is no resemblance between what you hear when the test is run and what you are seeing on the screen, Fig. 1.18.

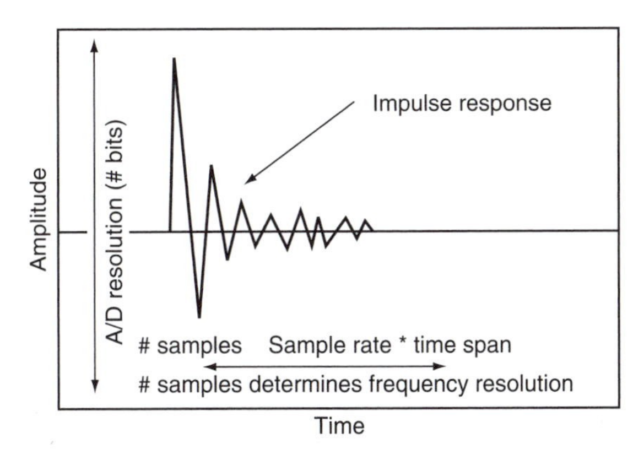

FIGURE 1.18 Many analyzers acquire the room response by digital sampling.

We can usually assume that the first energy arrival is from the loudspeaker itself, since any reflection would have to arrive later than the first wave front since it had to travel farther. Pre-arrivals can be caused by the acoustic wave propagating through a solid object, such as a ceiling or floor, and reradiating near the microphone. Such arrivals are very rare and usually quite low in level. In some cases a reflection may actually be louder than the direct arrival. This could be due to loudspeaker design or its placement relative to the mic location. It's up to the measurer to determine if this is normal for a given loudspeaker position/seating position. All loudspeakers will have some internal and external reflections that will arrive just after the first wave front. These are actually a part of the loudspeaker's response and can't be separated from the first wave front with a time window due to their close proximity without extreme compromises in frequency resolution. Such reflections are at least partially responsible for the characteristic sound of a loudspeaker. Studio monitor designers and studio control room designers go to great lengths to reduce the level of such reflections, yielding more accurate sound reproduction. Good system design practice is to place loudspeakers as far as possible from boundaries (at least at mid- and high frequencies). This will produce an initial time gap between the loudspeaker's response and the first reflections from the room. This gap is a good initial dividing point between the loudspeaker's response and the room's response, with the energy to the left of the dividing cursor being the response of the loudspeaker and the energy to the right the response of the room. The placement of this divider can form a time window by having the analyzer ignore everything later in time than the cursor setting. The time window size also determines the frequency resolution of the postprocessed data. In the frequency domain, improved resolution means a smaller number. For instance, 10 Hz resolution is better than 40 Hz resolution. Since time and frequency have an inverse relationship, the time window length required to observe 10 Hz will be much longer than the time window length required to resolve 40 Hz. The resolution can be estimated by $f = 1/T$, where T is the length of the time window in seconds. Since a frequency magnitude plot is made up of a number of data points connected by a line, another way to view the frequency resolution is that it is the number of Hz between the data points in a frequency domain display.

The method of determination of the time window length varies with different analyzers. Some allow a cursor to be placed anywhere on the data record, and the placement determines the frequency resolution of the spectrum determined by the window length. Others require that the measurer select the number of samples to be used to form the time window, which in turn determines the frequency resolution of the time window. The window can then be positioned at different places on the time domain plot to observe the spectral content of the energy within the window, Figs. 1.19, 1.20, and 1.21.

For instance, a one second total time (44,100 samples) could be divided into about twenty-two time windows of 2048 samples each (about 45 ms). Each window would allow the observation of

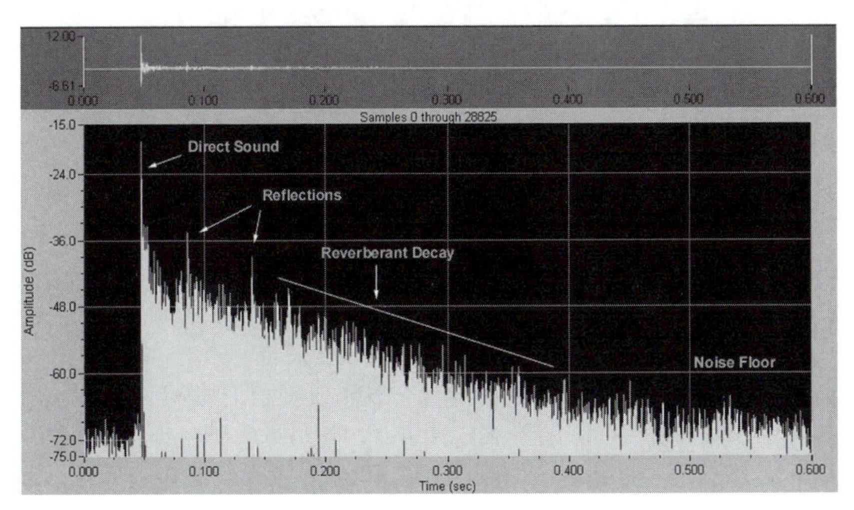

FIGURE 1.19 A room response showing the various sound fields that can exist in an enclosed space, SIA-SMAART.

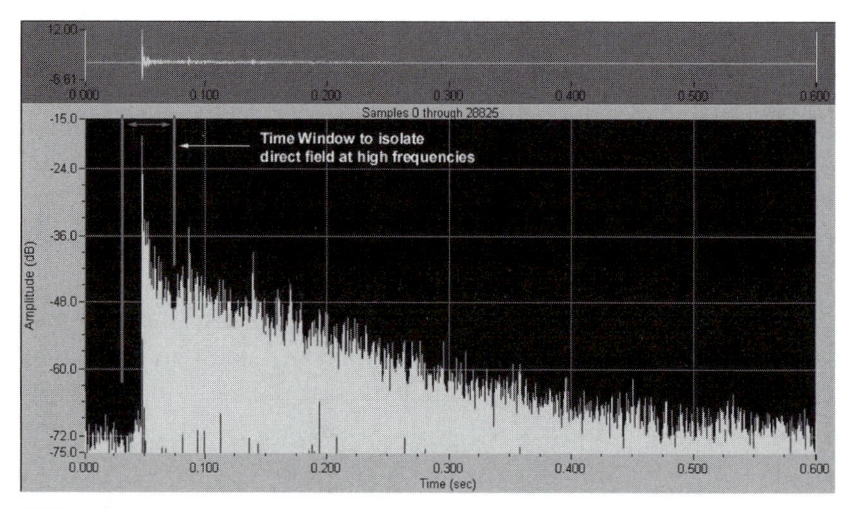

FIGURE 1.20 A time window can be used to isolate the loudspeaker's response from the room reflections.

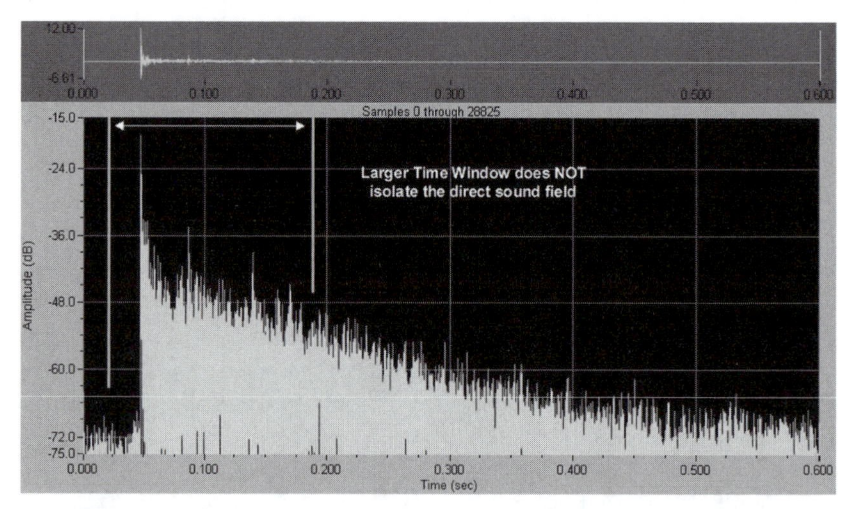

FIGURE 1.21 Increasing the length of the time window increases the frequency resolution, but lets more of the room into the measurement, SIA-SMAART.

the spectral content down to ($\frac{1}{45}$) × 1000 or 22 Hz. The windows can be overlapped and moved around to allow more precise selection of the time span to be observed. Displaying a number of these time windows in succession, each separated by a time offset, can form a 3D plot known as a waterfall.

1.3.5.9 Data Windows

There are some conditions that must be observed when placing cursors to define the time window. Ideally, we would like to place the cursor at a point on the time record where the energy is zero. A cursor placement that cuts off an energy arrival will produce a sharp rise or fall time that produces artifacts in the resultant calculated spectral response. Discontinuities in the time domain have broad spectral content in the frequency domain. A good example is a scratch on a phonograph record. The discontinuity formed by the scratch manifests itself as a broadband click during playback. If an otherwise smooth wheel has a discontinuity at one point, it would thump annoyingly when it was rolled on a smooth surface. Our measurement systems treat the data within the selected window as a continuously repeating event. The end of the event must line up with the beginning or a discontinuity occurs resulting in the generation of high-frequency artifacts called *spectral leakage*. In the same manner that a physical discontinuity in a phonograph record or wheel can be corrected by polishing, a discontinuity in a sampled time measurement can be remedied by tapering the energy at the beginning and end of the window to zero using a mathematical function. A number of data window shapes are available for performing the smoothing.

These include the Hann, Hamming, Blackman-Harris, and others. In the same way that a physical polishing process removes some good material from what is being rubbed, data windows remove some good data in the process of smoothing the discontinuity. Each window has a particular shape that leaves the data largely untouched at the center of the window but tapers it to varying degrees toward the edges. Half windows only smooth the data at the right edge of the time record while full windows taper both (start and stop) edges. Since all windows have side effects, there is no clear preference as to which one should be used. The Hann window provides a good compromise between time record truncation and data preservation. Figs. 1.22 and 1.23 show how a data window might be used to reduce spectral leakage.

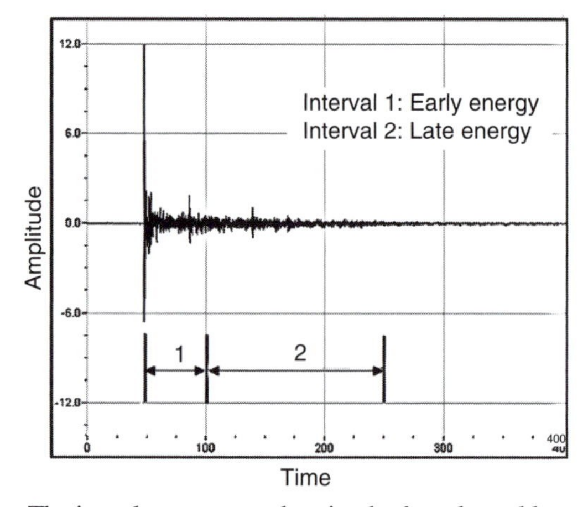

FIGURE 1.22 The impulse response showing both early and late energy arrivals.

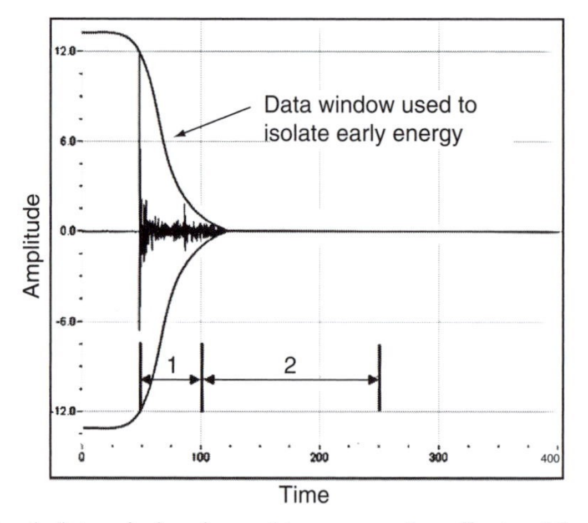

FIGURE 1.23 A data window is used to remove the effects of the later arrivals.

1.3.5.10 A Methodical Approach

Since there are an innumerable number of tests that can be performed on a system, it makes sense to establish a methodical and logical process for the measurement session. One such scenario may be as follows:

1. Determine the reason for and scope of the measurement session. What are you looking for? Can you hear it? Is it repeatable? Why do you need this information?

2. Determine what you are going to measure. Are you looking at the room or at the sound system? If it is the room, possibly the only meaningful measurements will be the overall decay time and the noise floor. If you are looking at the sound system, decide if you need to switch off or disconnect some loudspeakers. This may be essential to determine whether the individual components are working properly, or if an anomaly is the result of interaction between several components. "Divide and conquer" is the axiom.
3. Select the microphone position. I usually begin by looking at the on-axis response of the loudspeaker as measured from inside of critical distance. If multiple loudspeakers are on, turn all but one of them off prior to measuring. The microphone should be placed in the far free field of the loudspeaker as previously described. When measuring a loudspeaker's response, care should be taken to eliminate the effects of early reflections on the measured data, as these will generate acoustic comb filters that can mask the true response of the loudspeaker. In most cases the predominant offending surface will be the floor or other boundaries near the microphone and loudspeaker. These reflections can be reduced or eliminated by using a ground plane microphone placement, a tall microphone stand (when the loudspeaker is overhead), or some strategically placed absorption. I prefer the tall microphone stand for measuring installed systems with seating present since it works almost anywhere, regardless of the seating type. The idea is to intercept the sound on its way to a listener position, but before it can interact with the physical boundaries around that position. These will always be unique to that particular seat, so it is better to look at the free field response, as it is the common denominator to many listener seats.
4. Begin with the big picture. Measure an impulse response of the complete decay of the space. This yields an idea of the overall properties of the room/system and provides a good point of reference for zooming in to smaller time windows. Save this information for documentation purposes, as later you may wish to reopen the file for further processing.

5. Reduce the size of the time window to eliminate room reflections. Remember that you are trading off frequency resolution when truncating the time record, Fig. 1.24. Be certain to maintain sufficient resolution to allow adequate low-frequency detail. In some cases, it may be impossible to maintain a sufficiently long window to view low frequencies and at the same time eliminate the effects of reflections at higher frequencies, Fig. 1.25. In such cases, the investigator may wish to use a short window for looking at the high-frequency direct field, but a longer window for evaluating the woofer. Windows appropriate for each part of the spectrum can be used. Some measurement systems provide variable time windows, which allow low frequencies to be viewed in great detail (long time window) while still providing a semianechoic view (short time window) at high frequencies. There is evidence to support that this is how humans process sound information, making this method particularly interesting, Fig. 1.26.
6. Are other microphone positions necessary to characterize this loudspeaker? The off-axis response of some loudspeakers is very similar to the on-axis response, reducing the need to measure at

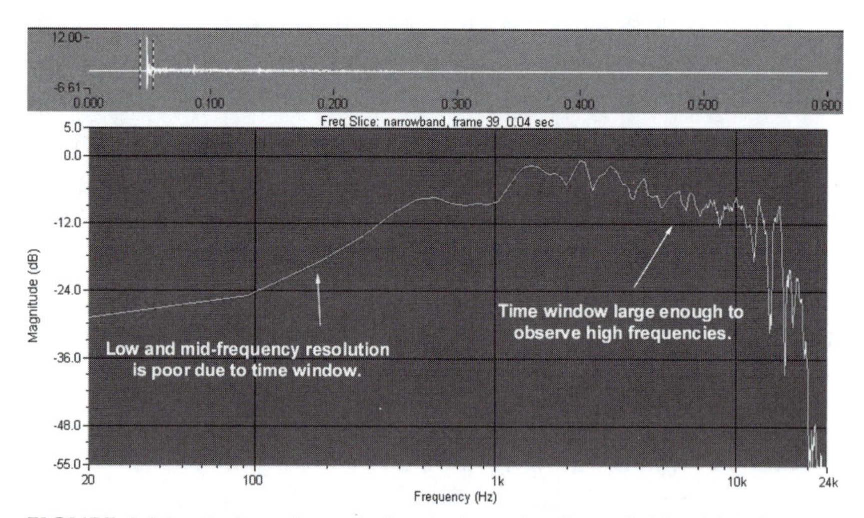

FIGURE 1.24 A short time window isolates the direct field at high frequencies at the expense of low-frequency resolution, SIA-SMAART.

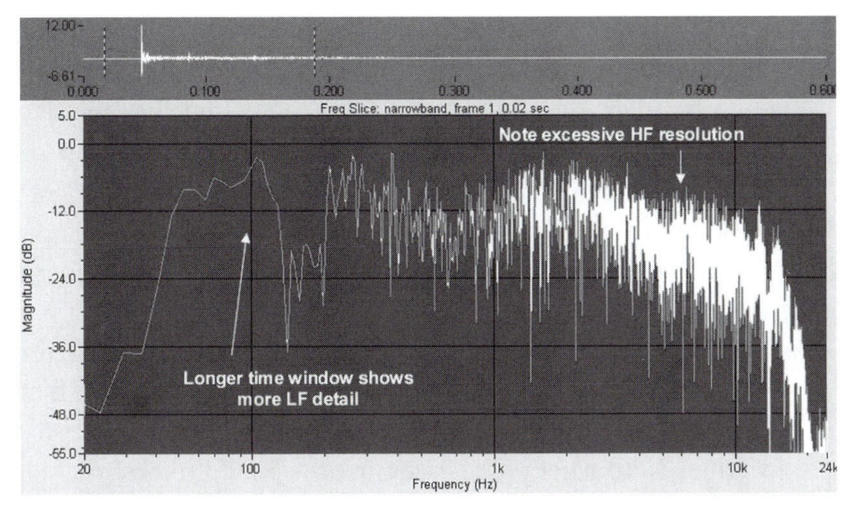

FIGURE 1.25 A long time window provides good low-frequency detail, SIA-SMAART.

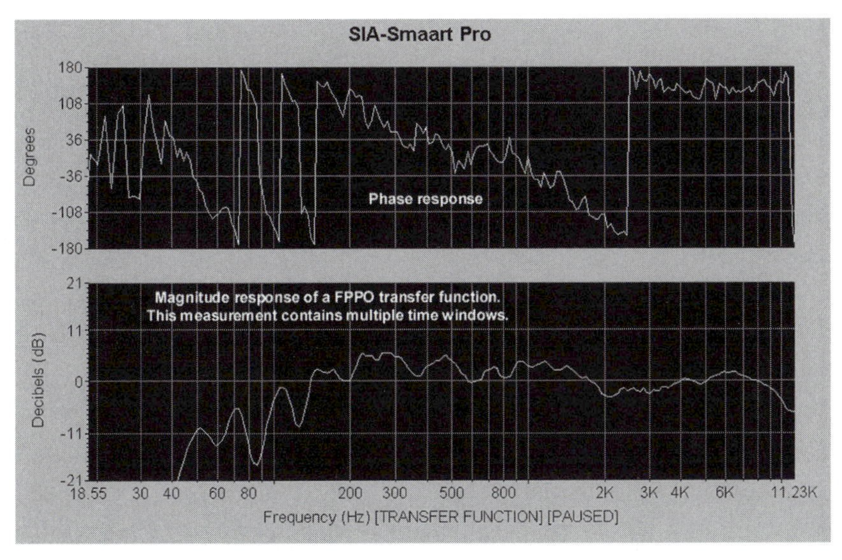

FIGURE 1.26 The time window increases with length as frequency decreases.

many angles. Other loudspeakers have very erratic responses, and a measurement at any one point around the loudspeaker may bear little resemblance to the response at other positions. This is a design issue, but one that must be considered by the measurer.

7. Once an accurate impulse response is measured, it can be postprocessed to yield information on spectral content, speech intelligibility, and music clarity. There are a number of metrics that can provide this information. These are interpretations of the measured data and generally correlate with subjective perception of the sound at that seat.
8. An often overlooked method of evaluating the impulse response is the use of convolution to encode it onto anechoic program material. An excellent freeware convolver called GratisVolver is available from www.catt.se. Listening to the IR can often reveal subtleties missed by the various metrics, as well as provide clues as to what post process must be used to observe the event of interest.

1.3.6 Human Perception

Useful measurement systems can measure the impulse response of a loudspeaker/room combination with great detail. Information regarding speech intelligibility and music clarity can be derived from the impulse response. In nearly all cases, this involves post processing the impulse response using one of several clarity measure metrics.

1.3.6.1 Percentage Articulation Loss of Consonants (%Alcons)

For speech, one such metric is the percentage articulation loss of consonants, or %Alcons. Though not in widespread use today, a look at it can provide insight into the requirements for good speech intelligibility. A %Alcons measurement begins with an impulse response, which is usually displayed as a log-squared response or ETC. Since the calculation essentially examines the ratio between early energy, late energy, and noise, the measurer must place cursors on the display to define these parameters. These cursors may be placed automatically by the measurement program. The result is weighted with regard to decay time, so this too must be defined by the measurer. Analyzers such as the TEF25™ and EASERA include best guess default placements based on the research of Peutz, Davis, and others, Fig. 1.27.

These placements were determined by correlating measured data with live listener scores in various acoustic environments, and represent a defined and orderly approach to achieving meaningful results that correlate with the perception of live listeners. The measurer is free to choose alternate cursor placements, but great care must be taken to be consistent. Also, alternate cursor placements make it difficult if not impossible to compare your results with those obtained by other measurers. In the default %Alcons placement, the early energy (direct sound field) includes the first major sound arrival and any energy arrivals within the next 7–10 ms. This forms a tight time span for the direct sound. Energy beyond this span is considered late energy and an impairment to communication. As one might guess, a later cursor placement yields better intelligibility scores, since more of the room response is being considered beneficial to intelligibility. As such, the default placement yields a worst-case scenario. The default placement considers the effects of the early-decay time (EDT) rather than the classical

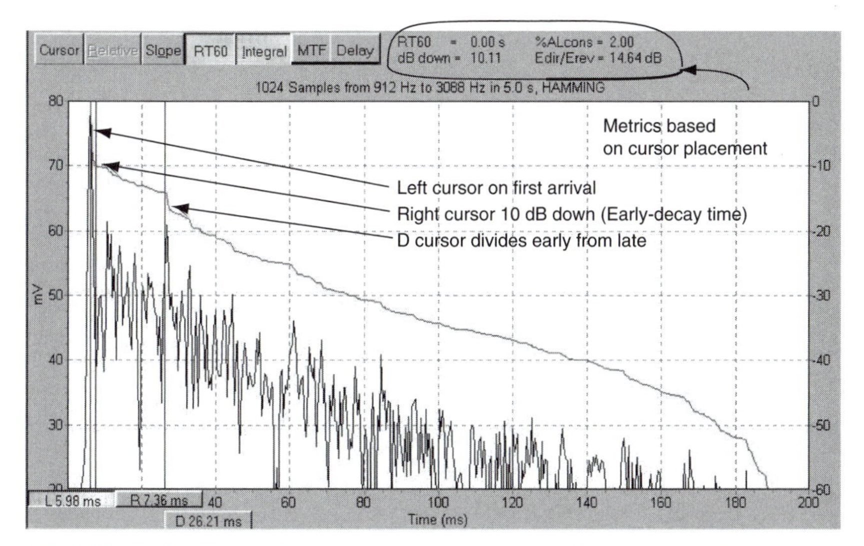

FIGURE 1.27 The ETC can be processed to yield an intelligibility score, TEF25.

T_{30} since short EDTs can yield good intelligibility, even in rooms with a long T_{30}. Again, the measurer is free to select an alternative cursor placement for determining the decay time used in the calculation, with the same caveats as placing the early-to-late dividing cursor. The %Alcons score is displayed instantly upon cursor placement and updates as the cursors are moved.

1.3.6.2 Speech Transmission Index (STI)

The STI can be calculated from the measured impulse response with a routine outlined by Schroeder and detailed by Becker in the reference. The STI is probably the most widely used contemporary measure of intelligibility. It is supported by virtually all measurement platforms, and some handheld analyzers are available for quick checks. In short, it is a number ranging from 0 to 1, with fair intelligibility centered at 0.5 on the scale. For more details on the Speech Transmission Index, see the chapter on speech intelligibility in this text.

1.3.7 Polarity

Good sound system installation practice dictates maintaining proper signal polarity from system input to system output. An audio signal waveform always swings above and below some reference point. In acoustics, this reference point is the ambient atmospheric pressure. In an electronic device, the reference is the 0 VA reference of the power supply (often called *signal ground*) in push-pull circuits or a fixed dc offset in class A circuits. Let's look at the acoustic situation first. An increase in the air pressure caused by a sound wave will produce an inward deflection of the diaphragm of a pressure microphone (the most common type) regardless of the microphone's orientation toward the source. This inward deflection should cause a positive-going voltage swing at the output of the microphone on pin 2 relative to pin 3, as well as at the output of each piece of equipment that the signal passes through. Ultimately the electrical signal will be applied to a loudspeaker, which should deflect outward (toward an axial listener) on the positive-going signal, producing an increase in the ambient atmospheric pressure. Think of the microphone diaphragm and loudspeaker diaphragm moving in tandem and you will have the picture. Since most sound reinforcement equipment uses bipolar power supplies (allowing the audio signal to swing positive and negative about a zero reference point), it is possible for signals to become inverted in polarity (flipped over). This causes a device to output a negative-going voltage when it is fed a positive-going voltage. If the loudspeaker is reverse-polarity from the microphone, an increase in sound pressure at the microphone (compression) will cause a decrease in pressure in front of the loudspeaker (rarefaction). Under some conditions, this can be extremely audible and destructive to sound quality. In other scenarios it can be irrelevant, but it is always good to check.

System installers should always check for proper polarity when installing the sound system. There are a number of methods, some simple and some complex. Let's deal with them in order of complexity, starting with the simplest and least costly method.

1.3.7.1 The Battery Test

Low-frequency loudspeakers can be tested using a standard 9 V battery. The battery has a positive and negative terminal, and the spacing between the terminals is just about right to fit across the terminals of most woofers. The loudspeaker cone will move outward when the battery is placed across the loudspeaker terminals with the battery positive connected to the loudspeaker positive. While this is one of the most accurate methods for testing polarity, it doesn't work for most electronic devices or high-frequency drivers. Even so, it's probably the least costly and most accurate way to test a woofer.

1.3.7.2 Polarity Testers

There are a number of commercially available polarity test sets in the audio marketplace. The set includes a sending device that outputs a test pulse, Fig. 1.28, through a small loudspeaker (for testing microphones) or an XLR connector (for testing electronic

FIGURE 1.28 A popular polarity test set.

devices), and a receiving device that collects the signal via an internal microphone (loudspeaker testing) or XLR input jack. A green light indicates correct polarity and a red light indicates reverse polarity. The receive unit should be placed at the system output (in front of the loudspeaker) while the send unit is systematically moved from device to device toward the system input. A polarity reversal will manifest itself by a red light on the receive unit.

1.3.7.3 Impulse Response Tests

The impulse response is perhaps the most fundamental of audio and acoustic measurements. The polarity of a loudspeaker or electronic device can be determined from observing its impulse response, Figs. 1.29 and 1.30. This is one of the few ways to test flown loudspeakers from a remote position. It is best to test the polarity of components of multiway loudspeakers individually, since all of the individual components may not be polarized the same. Filters in the signal path (i.e., active crossover network) make the results more difficult to interpret, so it may be necessary to carefully test a system component (e.g., woofer) full-range for definitive results. Be sure to return the crossover to its proper setting before continuing.

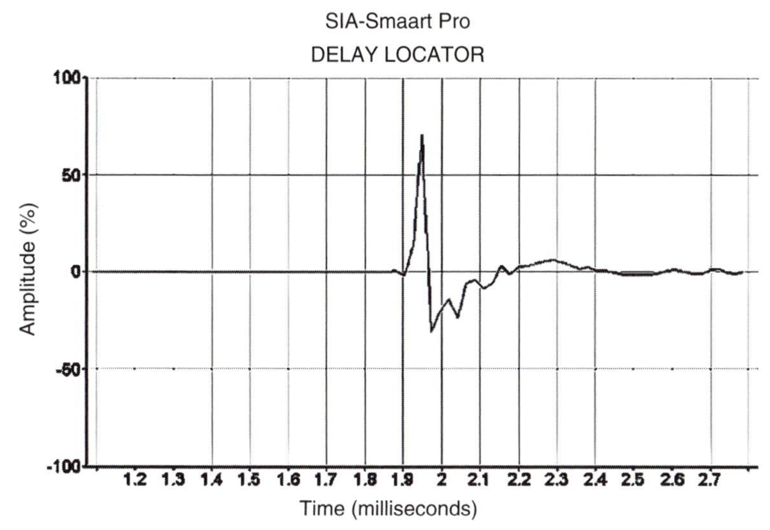

FIGURE 1.29 The impulse response of a transducer with correct polarity.

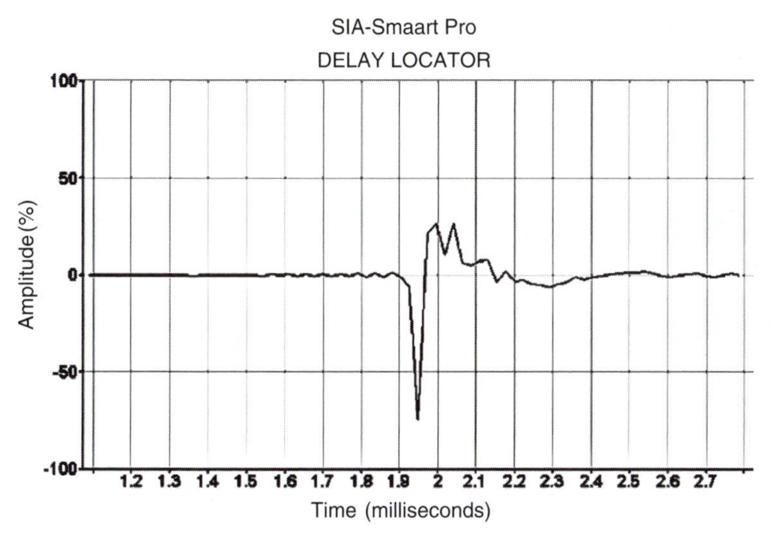

FIGURE 1.30 The impulse response of a reverse-polarity transducer.

1.4 CONCLUSION

The test and measurement of the sound reinforcement system are a vital part of the installation and diagnostic processes. The FFT and the analyzers that use it have revolutionized the measurement process, allowing sound practitioners to pick apart the system response and look at the response of the loudspeaker also. Powerful analyzers that were once beyond the reach of most technicians are readily available and affordable, and cost can no longer be used as an excuse for not measuring the system. The greatest investment by far is the time required to grasp the fundamentals of acoustics to allow interpretation of the data. Some of this information is general, and some of it is specific to certain measurement systems.

The acquisition of a measurement system is the first step in ascending the capability and credibility ladder. The next steps include acquiring proper instruction on its use by self-study or short course. The final and most important steps are the countless hours in the field required to correlate measured data with the hearing process. As proficiency in this area increases, the speed of execution, validity, and relevance of the measurements will increase also. While we can all learn how to make the measurements in a relatively short time span, the rest of our careers will be spent learning how to interpret what we are measuring.

REFERENCES

1. D. Davis and C. Davis, *Sound System Engineering*, Focal Press, 1997, Boston, MA.
2. The Heyser Anthology, Audio Engineering Society, NY, NY.

BIBLIOGRAPHY

Understanding Sound System/Room Interactions, Sam Berkow, Syn-Aud-Con Tech Topic, Volume 28, Number 2.

Joseph D'Appolitto, *Testing Loudspeakers*, Old Colony Sound Labs, Peterborough, NH.

Section 2

What's the Ear For? How to Protect It

Les Blomberg and Noland Lewis

2.1 WHAT'S THE EAR FOR?

An ear is for listening, and for the lucky few, listening to music is their job. But an ear is for much more—lose your hearing, and besides not hearing music, you lose your connection with other people. Hearing is the sense most related to learning and communication, and is the sense that connects you to ideas and other people. Helen Keller, who lost both her sight and hearing at a young age, said that hearing loss was the greater affliction for this reason.

To professionals in the music industry, their hearing is their livelihood. To be able to hear well is the basis for sound work. Protecting your hearing will determine whether you are still working in the industry when you are 64, or even whether you can still enjoy music, and it will determine whether you will hear your spouse and grandchildren then, too.

2.1.1 What Does Hearing Damage Sound Like?

Hearing loss is the most common preventable workplace injury. Ten million Americans have noise-induced hearing loss. Ears can be easily damaged, resulting in partial or complete deafness or persistent ringing in the ears.

Hearing loss isn't necessarily quiet. It can be a maddening, aggravating buzz or ringing in the ear, called *tinnitus*. Or it may result in a loss of hearing ability, the ability to hear softer sounds at a particular frequency. The threshold of hearing (the softest sounds that are audible for each frequency) increases as hearing loss progresses. Changes in this threshold can either be a temporary threshold shift (TTS) or a permanent threshold shift (PTS). Often these changes occur in the higher frequencies of 3000 to 6000 Hz, with a notch or significant reduction in hearing ability often around 4000 Hz.

A single exposure to short-duration, extreme loud noise or repeated and prolonged exposure to loud noises are the two most common causes of hearing loss. Examples of the first might be exposure to noise from discharging firearms, while the second might be the cumulative effect of working in a noisy environment such as a factory or in loud concert venues. Some antibiotics, drugs, and chemicals can also cause permanent injury.

Hearing damage isn't the only health effect of noise. Workers in noisy workplaces have shown a higher likelihood of heart disease and heart attacks. Numerous other stress-related effects have been documented, including studies that have shown that women in noisy environments tend to gain weight.

2.2 HOW LOUD IS TOO LOUD? OSHA, NIOSH, EPA, WHO

As in other industries, workers in the sound industry are covered by the occupational noise exposure standard found in the Code of Federal Regulations (29 CFR 1910.95). Occupational Safety and Health (OSHA) regulation requires that workers' exposures not exceed those in Table 2.1.

TABLE 2.1 Permissible Noise Exposures

Duration per Day, Hours	Sound Level dBA Slow Response
8	90
6	92
4	95
3	97
2	100
1½	102
1	105
½	110
¼ or less	115

Noise levels are measured with a sound level meter or dosimeter (a sound level meter worn on the employee) that can automatically determine the average noise level. Often, noise levels are represented in terms of a daily dose. For example, a person who was exposed to an average level of 90 dBA for four hours would have received a 50% dose, or half of her allowable exposure.

Administrative controls—such as the boss saying, "Don't work in noisy areas, or do so for only short times," and/or engineering controls—such as quieter machines—are required to limit exposure. Hearing protection may also be used, although it is not the preferred method. Moreover, the regulation requires that, for employees whose exposure may equal or exceed an 8-hour time-weighted average of 85 dB, the employer shall develop and implement a monitoring program in which employees receive an annual hearing test. The testing must be provided for free to the employee. The employer is also required to provide a selection of hearing protectors and take other measures to protect the worker.

Compliance by employers with the OSHA regulations, as well as enforcement of the regulation, is quite variable, and often it is only in response to requests from employees. It is quite possible that professionals in the field have never had an employer-sponsored hearing test, and are not participating in a hearing conservation program as required.

Unfortunately, OSHA's regulations are among the least protective of any developed nation's hearing protections standards. Scientists and OSHA itself have known for more than a quarter-century that nearly a third of the population exposed to OSHA-permitted noise levels over their lifetime will suffer substantial hearing loss: see Table 2.2. As a result, the National Institute of Occupational Safety and Health (NIOSH), a branch of the Centers for Disease Control and Prevention (CDC), has recommended an 85 dB standard as shown in Table 2.3. Nevertheless, NIOSH recognizes that approximately 10% of the population exposed to the lower recommended level will still develop hearing loss.

Table 2.3 compares the permissible or recommended daily exposure times for noises of various levels. The table is complicated but instructive. The first three columns represent the recommendations of the Environmental Protection Agency (EPA) and World Health Organization (WHO) and starts with the recommendation that the 8-hour average of noise exposure not exceed 75 dBA. The time of exposure is reduced by half for each 3 dBA that is added; a 4-hour exposure is 78 dBA, and a 2-hour exposure is 81 dBA. This is called a 3 dB exchange rate, and is justified on the principle that a 3 dB increase is a doubling of the energy received by the

TABLE 2.2 NIOSH's 1997 Study of Estimating Excess Risk of Material Hearing Impairment

Average Exposure Level–dBA	Risk of Hearing Loss Depending on the Definition of Hearing Loss Used
90 (OSHA)	25–32%
85 (NIOSH)	8–14%
80	1–5%

While 25–32% of the population will suffer substantial hearing loss at OSHA permitted levels, everyone would suffer some hearing damage.

TABLE 2.3 EPA, WHO, NIOSH, and OSHA Recommended Decibel Standards

	EPA and WHO			NIOSH			OSHA	
dBA	Hours	Mins.	Secs.	Hours	Mins.	Secs.	Hours	Mins.
75	8							
76								
77								
78	4							
79								
80								
81	2							
82								
83								
84	1							
85				8				
86								
87		30						
88				4				
89								
90		15					8	
91				2				
92								
93		7	30					
94				1				
95							4	
96		3	45					
97					30			
98								
99		1	53					
100					15		2	

TABLE 2.3 EPA, WHO, NIOSH, and OSHA Recommended Decibel Standards—Contd.

	EPA and WHO			NIOSH			OSHA	
dBA	Hours	Mins.	Secs.	Hours	Mins.	Secs.	Hours	Mins.
101								
102			56					
103					7	30		
104								
105			28				1	60
106					3	45		
107								
108			14					
109					1	53		
110							0.5	30
111			7					
112						56		
113								
114			4					
115						28	0.25	15

ear, and therefore exposure time ought to be cut in half. The EPA and WHO recommendations can be thought of as safe exposure levels. The NIOSH recommendations in the next three columns represent an increased level of risk of hearing loss and are not protective for approximately 10% of the population. NIOSH uses a 3 dB exchange rate, but the 8-hour exposure is 10 dB higher than EPA—that is, 85 dBA. Finally, the OSHA limits are in the last two columns. OSHA uses a 5 dB exchange rate, which results in much longer exposure times at higher noise levels, and the 8-hour exposure is 90 dBA. Between 25 and 32% of people exposed to OSHA-permitted levels will experience significant hearing loss over a lifetime of exposure. It is important to note that everyone exposed to the OSHA-permitted levels over their lifetime will experience some hearing loss.

Remember that each of these recommendations assumes that one is accounting for all of the noise exposure for the day. Someone who works in a noisy environment, and then goes home and uses power tools or lawn equipment is further increasing their risk and exposure.

The U.S. Environmental Protection Agency (EPA) and the World Health Organization (WHO) have recommended a 75 dB limit, as shown in Table 2.3, as a safe exposure with minimal risk of hearing loss. The WHO goes on to recommend that exposure such as encountered at a rock concert be limited to four times per year.

2.3 INDICATORS OF HEARING DAMAGE

There are several indicators of hearing damage. Since the damage is both often slow to manifest itself and progressive, the most important indicators are the ones that can be identified before permanent hearing damage has occurred.

The first and most obvious indicator is exceeding the EPA and WHO safe noise levels. After 8 hours of noise exposure, the risk of suffering hearing loss also increases.

Exceeding the safe levels by, for example, working at OSHA-permitted noise levels doesn't necessarily mean you will suffer substantial hearing loss; some people will suffer substantial loss, but everyone will suffer some level of hearing damage. The problem is that there is no way to know if you are in the one quarter to one third of the population who will suffer substantial hearing loss at a 90 dBA level or the two thirds to three quarters of the population who will lose less—at least, not until it is too late and the damage has occurred. Of course, by greatly exceeding OSHA limits, you can be assured that you will have significant hearing loss.

There are two types of temporary hearing damage that are good indicators that permanent damage will occur if exposure continues. The first is tinnitus, a temporary ringing in the ears following a loud or prolonged noise exposure. Work that induces tinnitus is clearly too loud, and steps should immediately be taken to limit exposure in the future.

The second type of temporary damage that is a useful indicator of potential permanent damage is a temporary threshold shift (TTS). Temporary changes in the threshold of hearing (the softest sounds that are audible for each frequency) are a very good indicator that continued noise exposure could lead to permanent hearing loss. Although ways to detect TTS without costly equipment are now being developed, the subjective experience of your hearing sounding different after noise exposure currently provides the best indication of problems.

It is important to remember that the absence of either of these indicators does not mean you will not suffer hearing loss. The presence of either is a good indication that noise exposure is too great.

Regular hearing tests can't detect changes in hearing before they become permanent, but if frequent enough, they can detect changes before they become severe. It is particularly important, therefore, that people exposed to loud noises receive regular hearing tests.

Finally, there are often indicators that serious hearing damage has occurred, such as difficulties understanding people in crowded, noisy situations (loud restaurants, for example), the need to say "What?" frequently, or asking people to repeat themselves. Often it is not the person with the hearing loss, but rather others around him or her, who are the first to recognize these problems due to the slow changes to hearing ability and the denial that often accompanies them. While it is impossible to reverse hearing damage, hearing loss can be mitigated somewhat by the use of hearing aids, and further damage can be prevented. It is important to remember that just because you have damaged your hearing doesn't mean you can't still make it much worse.

2.4 PROTECTING YOUR HEARING

Protecting your hearing is reasonably straightforward: avoid exposure to loud sounds for extended periods of time. This can be accomplished by either turning down the volume or preventing the full energy of the sound from reaching your ears.

There are several strategies for protecting your hearing if you believe or determine that your exposure exceeds safe levels. As

Table 2.3 indicates, you can reduce *the noise level* or reduce *the exposure time*, or both.

While reducing exposure time is straightforward, it is not always possible, in which case turning down the volume by using quieter equipment, maintaining a greater distance from the noise source, using barriers or noise-absorbing materials, or utilizing hearing protection (either earplugs or noise-cancelling headphones or both) are required.

Typical earplugs or headphones are often criticized for changing the sound and hindering communication. Hearing protection in general is far better at reducing noise in the higher frequencies than the lower frequencies, so typical hearing protection significantly changes the sound the wearer is hearing. Consonant sounds in speech occur in the frequencies that are more greatly attenuated by some hearing protectors.

There are, however, a number of hearing protection devices designed to reduce noise levels in all frequencies equally. Often referred to as musician's earplugs, these can come in inexpensive models or custom-molded models. The advantage of a flat or linear attenuation of noise across all frequencies is that the only change to the sound is a reduction in noise level.

2.4.1 Protecting Concert-Goers and Other Listeners

Ears are for listening, and when it comes to musical performances, there are often many ears present in the audience. Concert-goers, like music professionals, are at risk of hearing loss. Loud music is exciting; that is the physiology of loud. It gives us a shot of adrenaline. Also, more neurons are firing in our brain and our chest is resonating with the low-frequency sounds.

When humans evolved, the world was much quieter than it is today. Infrequent thunder was about the extent of loud noise. Hearing evolved to be a very important sense with respect to our survival, keeping us informed about the changing conditions of our environment. Noise alerts us, because if it didn't wake our forebears up when trouble entered the camp, they might not live long enough to create descendants. Noise is an important warning device—think of a child's cry or a shout. During most of human history, when it was loud, trouble was involved. Physiologically, loud noises give us a shot of adrenaline, gearing us up to either fight or flee. Today, while neither fight nor flight is an appropriate response to loud noise, we still receive that shot of adrenaline. This is the reason for the popularity of loud movie soundtracks, loud exercise gyms, and loud music. It adds excitement and energy to activities. But it is also the reason for the stress-related effects of noise.

There is great incentive to turn up the volume, especially since the consequences are often not experienced until years later when the extent of hearing damage becomes apparent. People come to concert venues for excitement, not to be bored, and they come willingly; in fact, they pay to suffer whatever damage might be caused. Still, it is not a well-informed decision, and often minors are in the audience. But mostly, high volume isn't necessary. The desired physiological responses occur at lower noise levels. Moreover, it makes little sense for an industry to degrade the experience of listening to music in the future for whatever marginal gain comes from turning it up a few more decibels now.

Fortunately, even small gestures to lower noise levels have noticeable impacts. Because every 3 dB decrease halves exposure, small decreases in sound pressure level can vastly increase public safety.

2.4.2 Protecting the Community

Noise can spill over from a venue into the community. The term *noise* has two very different meanings. When discussing hearing loss, noise refers to a sound that is loud enough to risk hearing damage. In a community setting, noise is aural litter. It is audible trash. Noise is to the soundscape as litter is to the landscape. When noise spills over into the community, it is the aural equivalent of throwing McDonald's wrappers onto someone else's property.

When noise reaches the community, often it has lost its higher-frequency content, as that is more easily attenuated by buildings, barriers, and even the atmosphere. What is often left is the bass sound.

Solutions to community noise problems are as numerous as the problems themselves, and usually require the expertise of architectural acousticians. In general, carefully aimed distributed speaker systems are better than large stacks for outdoor venues. Barriers can help, but not in all environmental conditions, and their effectiveness tends to be limited to nearer neighbors. Moreover, barriers need to be well designed, with no gaps.

Indoor walls with higher sound transmission class (STC) ratings are better than ones with lower ratings. STC ratings, however, do not address low-frequency sounds that are most problematic in community noise situations, so professional advice is important when seeking to design better spaces or remedy problems.

Windows and doors are particularly problematic, as even these small openings can negate the effects of very well-soundproofed buildings. They also tend to be the weakest point, even when shut.

Sound absorption is useful for reducing transmission through walls, but in general, decoupling the interior and exterior so that the sound vibrations that hit the interior wall do not cause the exterior wall to vibrate and reradiate the noise is more effective. There are numerous products available to achieve both decoupling and sound absorption.

Often, however, employing these techniques is not an option for the sound engineer. In that case, controlling sound pressure levels and low-frequency levels are the best solution.

2.5 TOO MUCH OF A GOOD THING

In today's world, noise represents one of the more serious pollutants. The Romans had a word for it. Fig. 2.1. Some are the by-products of our society such as lawn mowers, jackhammers, traffic, and public transportation.

We deliberately subject ourselves to a Pandora's box of sounds that threaten not only our hearing but our general health. Personal devices like MP3 players, car stereos, or home theaters are sources we can control, yet many remain oblivious to their impact, Fig. 2.2. In the public domain, clubs, churches, auditoriums, amphithe-aters, and

FIGURE 2.1 Derivation of noise. Courtesy ACO Pacific.

FIGURE 2.2 Acceptance of some "noise" is often based on preference. Courtesy ACO Pacific.

stadiums are part of the myriad of potential threats to hearing health. From a nuisance to a serious health risk, these sources impact attendees, employees, and neighbors alike. As pointed out previously, levels of 105 dBA for 1 hour or less may result in serious and permanent hearing damage. Recent studies have shown that other factors such as smoking, drugs of all types, and overall ill-health appear to accelerate the process.

High sound levels are just part of the problem. Sound does not stop at the property line. Neighbors and neighborhoods are affected. Numerous studies have shown persistent levels of *noise* affect sleeping patterns, even increase the potential for heart disease. Studies by Johns Hopkins have shown hospital noise impacts patients in the neonatal wards and other patients' recovery time.

Communities all over the world have enacted various forms of noise ordinances. Some address *noise* based on the annoyance factor. Others specify noise limits with sound pressure level (SPL), time of day, and day of the week regulations. Unfortunately, noise, and sound itself, is a transient event. Consequently, enforcement and compliance are often very difficult to accomplish.

2.5.1 A Compliance and Enforcement Tool

There are various tools to monitor noise. One very useful tool is the SLARM™ by ACO Pacific. The following will use the SLARM™ to explain the importance of noise-monitoring test gear. The SLARM™ tool was developed to meet the needs of the noise abatement market. The SLARM™ performs both compliance and enforcement roles, offering accurate measurement, alarm functions, and very important history.

For the business owner dealing with neighborhood complaints, the SLARM™ provides a positive indication of SPL limits—permitting employees to control the levels or even turn off the sound. The History function offers a positive indication of compliance.

On the enforcement side, no longer does enforcement have to deal with finger-pointing complaints. They now may be addressed hours or days after the event, and resolved. There is also the *uniform effect*. Police pull up armed with a sound level meter (SLM) and the volume goes down. Businesses now can demonstrate compliance. Yes, it is an oversimplification, the concept works. Agreements are worked out. Peace and quiet return to the neighborhood.

2.5.1.1 The SLARMSolution™

The SLARM™ (**S**ound **L**evel **A**larm and **M**onitor) is a package of three basic subsystems in a single standalone device:

1. A sound level meter designed to meet or exceed Type 1 specifications
2. Programmable threshold detectors providing either SPL or Leq alarm indications
3. Monitor—a data recorder storing SPL data, and Led values for about 3 weeks on a rolling basis, as well as logging unique Alarm events, scheduled threshold changes, maintenance events, and calibration information

The SLARM™ may operate standalone. A PC is not required for normal Alarm operation. The data is maintained using flash and ferro-ram devices.

The SLARM™ provides USB and serial connectivity. It may be connected directly to a PC or via optional accessories directly to an Ethernet or radio link such as Bluetooth™.

PC operation is in conjunction with the included SLARMSoft™ software package.

2.5.1.2 SLARMSoft™ Software Suite

***SLARMWatch*™.** A package with password-protected setup, calibration, downloading, display, and clearing of the SLARM's™ SPL history. The history data may be saved and imported for later review and analysis, Fig. 2.3.

***SLARMAnalysis*™.** Part of SLARMWatch™ provides tools for the advanced user to review the SLARM™ history files. SLARMWatch™ allows saving and storage of this file for later review and analysis. SLARMAnalysis™ provides Leq, Dose and other calculations with user parameters, Fig. 2.4.

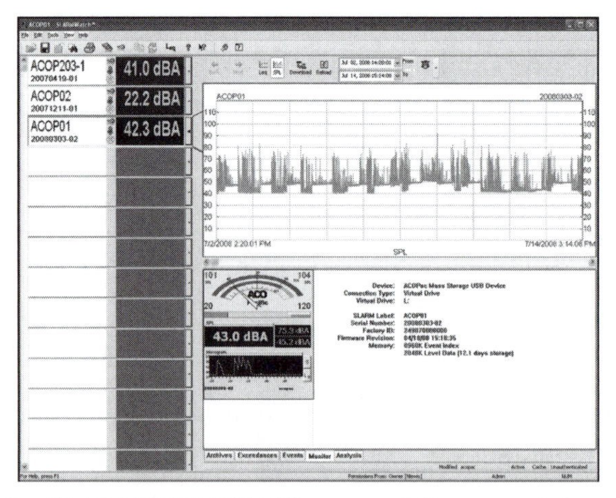

FIGURE 2.3 SLARMWATCH™ History and Events and three SLARM™ displays. Courtesy ACO Pacific.

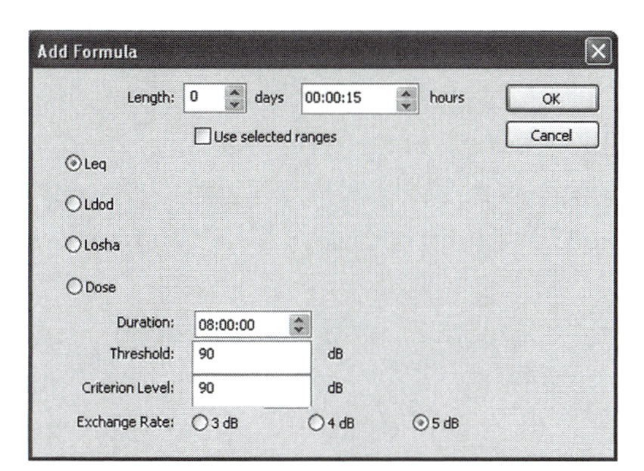

FIGURE 2.4 SLARMAnalysis™ Panel Courtesy, ACO Pacific.

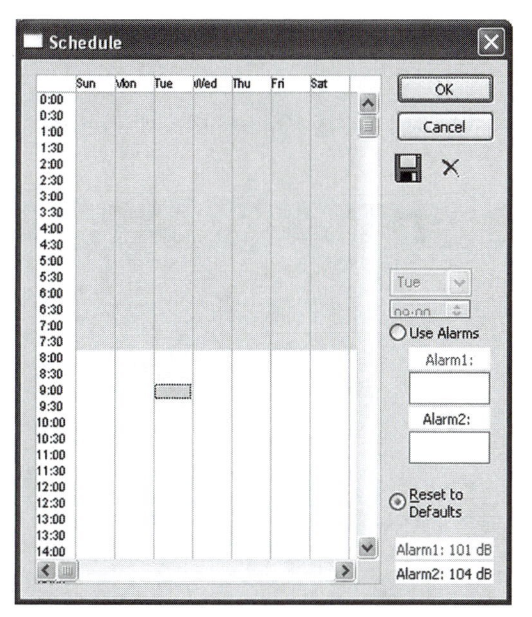

FIGURE 2.5 SLARMScheduler™ Panel. Thresholds may be individually set for each Alarm over a 24-hour, 7-day period. Courtesy ACO Pacific.

***SLARMScheduler*™.** Part of the SLARMWatch™ package, allows 24/7 setting of the Alarm thresholds. This permits time of day and day of the week adjustments to meet the needs of the community, Fig. 2.5.

***WinSLARM*™.** A display of SPL, Leqs, Range, and Alarm settings with digital, analog bar graph, and meter displays, as well as a Histogram window that provides a 25 second view of recent SPL on a continuous basis. The WinSlarm™ display may be sized permitting single or multiple SLARMS™ to be shown, Fig. 2.6.

***SLARMAlarm*™.** Operates independently from SLARMWatch™. The package monitors SLARMS™ providing digital display of SPL and Leqs values while also offering SMS, text, and email messaging of Alarm events via an Internet connection from the PC, Fig. 2.7.

***SLARMNet*™.** The SLARM™ and the SLARMSoft™ package allow multiple SLARMS™ to be connected to a network providing real-time data with alarm indications to multiple locations.

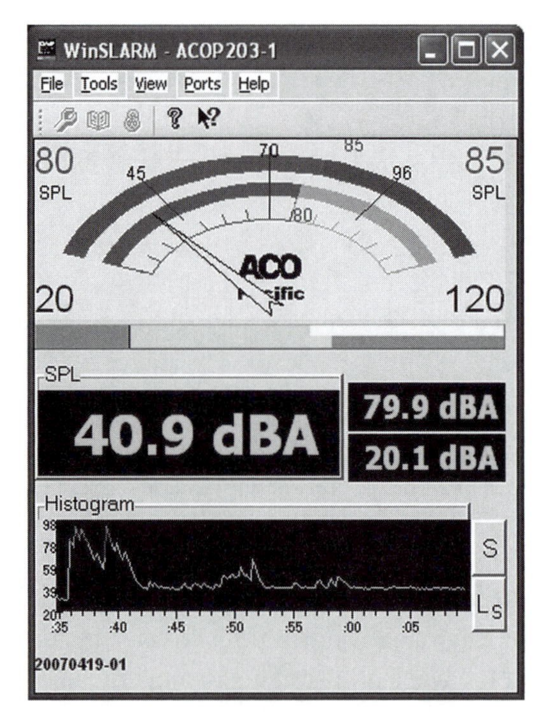

FIGURE 2.6 WinSLARM™ display provides a real-time look at SPL, Leq Thresholds, and recent events. Courtesy ACO Pacific.

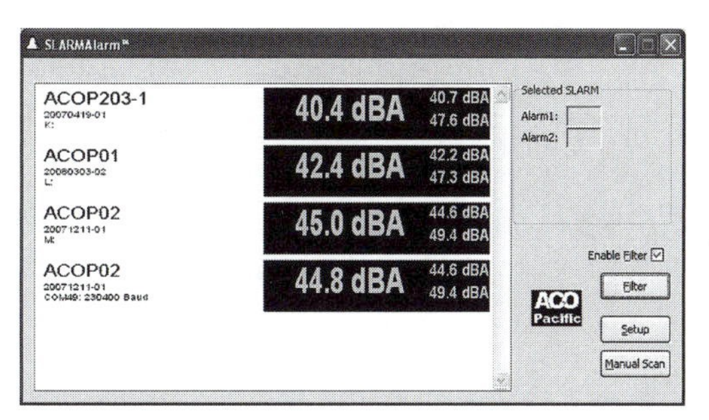

FIGURE 2.7 SLARMAlarm™ display with three SLARMS™. Note: ACOP2 has both USB and Ethernet (via a serial adaptor) connections. Courtesy ACO Pacific.

2.5.1.3 SLARM™ Operation

The SLARM™ operates in the following manner, Fig. 2.8.

The Microphone and Microphone Preamplifier. The 7052/4052 microphone and preamplifier are supplied with the SLARM™ system. The 7052 is a Type 1.5™ ½ inch free-field measurement microphone featuring a titanium diaphragm. The microphone has a frequency response from <5 Hz to 22 kHz and an output level of 22 mV/Pa (–33 dBV/Pa). The 4052 preamplifier is powered from 12 Vdc supplied by the SLARM™ and has a response <20 Hz to >100 kHz. Together they permit measurements approaching 20 dBA. The MK224 electret capsule is available, offering 8 Hz to 20 kHz response, and 50 mV/Pa (–26 dBV/Pa) performance providing a lower noise floor. The diaphragm is quartz-coated nickel.

The Preamplifier (Gain Stage). A low noise gain stage is located after the microphone input. This stage performs two tasks. The first limits the low-frequency input to just under 10 Hz. This reduces low-frequency interference from wind or doors slamming, things we do not hear due to the roll-off of our hearing below 20 Hz. The gain of this stage is controlled by the microcontroller providing two 100 dB measurement ranges 20 to 120 dB and 40 to 140 dBSPL. Most measurements are performed with the 20 to 120 dBSPL ranges. Custom ranges to >170 dBSPL are available

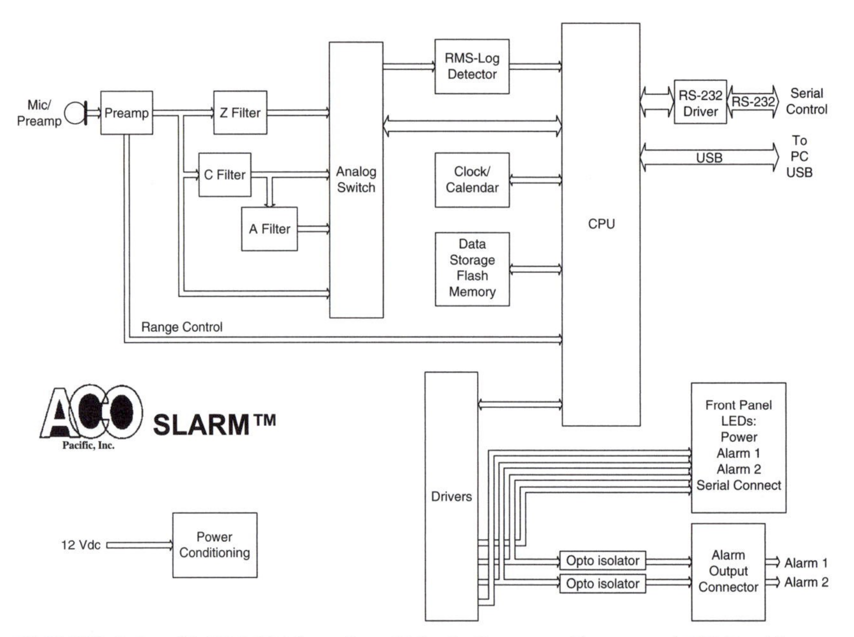

FIGURE 2.8 SLARM™ functional block diagram. Courtesy ACO Pacific.

as options. The output of the gain stage is supplied to three analog filter stages “A,” “C” and “Z” (Linear).

Analog A- and C-Weighted Filters. The gain stage is fed to the C-weighted filter. C-weighted filters have a –3 dB response limit of 31.5 Hz to 8 kHz. C-weighted filters are very useful when resolving issues with low frequencies found in music and industrial applications. The output of the C-weighted filter is connected to both the analog switch providing filter selection and the input of the A-weighted element of the filter system. Sound levels measured with the C-weighted filter are designated as dBC (dBSPL C-weighted).

The A-weighted response is commonly found in industrial and community noise ordinances. A-weighting rolls off low-frequency sounds. Relative to 1 kHz, the roll-off is –19.4 dB at 100 Hz (a factor of 1:10) and –39.14 at 31.5 Hz (a factor of 1:100). The A response significantly deemphasizes low-frequency sounds. Sound levels measured with the A-weighted filter are designated as dBA (dBSPL A-weighted). The output of the A-weighted filter is sent to the analog switch.

Analog Z-Weighting (Linear) Filter. The Z designation basically means the electrical output of the microphone is not weighted. The SLARM™ Z-weighting response is 2 Hz to >100 kHz. The response of the system is essentially defined by the response of the microphone and preamp. Z-weighting is useful where measurements of frequency response are desired, or where low or high frequencies are important. Remember the microphone response determines the response. Sound levels measured with the Z-weighted filter are designated as dBZ (dBSPL Z-weighted).

Analog Switch. The outputs of the A-, C-, and Z-weighted filters connect to the analog switch. The switch is controlled by the microcontroller. The selection of the desired filter is done at setup using the utilities found in SLARMWatch™.

Selection of the filter as with the other SLARM™ settings is password protected. Permission must be assigned to the user by the administrator before selection is possible. This is essential to minimize the possibility of someone changing measurement profiles that may result in improper Alarm activation or inaccurate measurements.

RMS Detection and LOG Conversion. The output of the analog switch goes to the RMS detection and Logarithmic conversion section of the SLARM™. The RMS detector is a true RMS detector able to handle crest factors of 5–10. This is different from an averaging detector set up provide rms values from sine wave (low crest factor) inputs. The response of the detector exceeds the response limits of the SLARM™.

The output of the RMS detector is fed to the Log (Logarithmic) converter. A logarithmic conversion range of over 100 dB is obtained. The logarithmic output then goes to the A/D section of the microcontroller.

Microcontroller. The microcontroller is the digital heart of the SLARM™. A microcontroller (MCU) does all the internal calculations and system maintenance.

SPL, Leq. The digital data from the internal A/D is converted by the MCU to supply dBSPL, and Leq values for both storage in the on-board flash memory and inclusion in the data stream supplied to the USB and serial ports. These are complex mathematical calculations involving log and anti-log conversation and averaging.

The SPL values are converted to a rolling average. The results are sent to the on-board flash memory that maintains a rolling period of about 2 to 3 weeks.

Leq generation in the SLARM™ involves two independent calculations with two programmable periods. A set of complex calculations generates the two Leq values.

Thresholds and Alarms. The results of the Averaging and Leq calculations are compared by the microcontroller with the Threshold levels stored in the on-board ferro-ram. Threshold levels and types—SPL or Leq—are set using the Settings tools provided in SLARMWatch™. These thresholds are updated by the SLARMScheduler™ routine.

If the programmed threshold limits are exceeded, the microcontroller generates an output to an external driver IC. The IC decodes the value supplied by the microcontroller, lighting the correct front panel Alarm LED, and also activating an opto-isolator switch. The opto-switch contacts are phototransistors. The transistor turns on when the opto-isolator LED is activated. The result—a contact closure signaling the outside world of the Alarm.

Real-time Clock. The SLARM™ has an on-board real-time clock. Operating from an internal lithium cell, the real-time clock timestamps all of the recorded history, event logging, and also controls the SLARMScheduler™ operation. The Settings panel in SLARMWatch™ allows user synchronization with a PC.

Communicating with the Outside World. SLARM™ may be operated standalone (without a PC). The SLARM™ provides both USB 2.0 and RS232 serial connections. The USB port is controlled by the microcontroller and provides full access to the SLARM settings, History flash memory, and firmware update capability.

The RS232 is a fully compliant serial port capable of up to 230 k Baud. The serial port may be used to monitor the data stream from the SLARM™. The serial port may also be used to control the SLARM™ settings.

Ethernet and Beyond. Utilizing the wide variety of after-market accessories available, the USB and Serial ports of the SLARM™ may be connected to the Ethernet and Internet. RF links like Bluetooth® and WiFi are also possible. Some accessories will permit the SLARM™ to become an Internet accessory without a PC, permitting remote access from around the world.

The SLARMSoft™ package permits the monitoring of multiple SLARMS™ through the SLARMNet™. The SLARMAlarm™ software not only provides a simple digital display of multiple SLARMS™, but also permits transmission of SMS, text, and email of Alarm events. This transmission provides the SLARM™ ID, Time, Type, and Level information in a short message. The world is wired.

History. The on-board flash and ferro-ram memories save measurements, events, settings, user access, and the SLARM™ Label. The SLARM™ updates the flash memory every second. SPL/Leq data storage is on a rolling 2 to 3 week basis. Alarm events, user access, and setting changes are also logged. These may be downloaded, displayed, and analyzed using features found in SLARMWatch™.

2.5.1.3.1 Applications

SLARM™ applications are virtually unlimited. Day–to–day applications are many. Day care centers, hospitals, classrooms, offices, clubs, rehearsal halls, auditoriums, amphitheaters, concert halls, churches, health clubs, and broadcast facilities are among the locations benefiting from sound level monitoring. Industrial and community environments include machine shops, assembly lines, warehouses, marshaling yards, construction sites and local enforcement of community noise ordinances. The following are examples of recent SLARMSolution™.

A Healthy Solution. Located in an older building with a lot of flanking problems, the neighbors of a small women's health club were complaining about the music used with the exercise routines. Negotiations were at a standstill until measurements were made.

Music levels were measured in the health club and a mutually acceptable level established. A SLARM™ (operating standalone—no PC) was installed to monitor the sound system and a custom control accessory developed to the customer's specifications. If the desired SPL limits were exceeded for a specific period of time, the SLARM™ disabled the sound system, requiring a manual reset. The result: a Healthy Solution.

Making a Dam Site Safer. A SLARM™ (operating standalone—no PC) combined with an Outdoor Microphone assembly (ODM) located 300 ft away, monitors the 140+ dBSPL of a gate warning horn. The operator, over 100 miles away, controls the flood gates of the dam, triggering the horn. The PLC controls the gate operation and monitors power to the horn but not the acoustic output. The SLARMSolution™ monitors the sound level from the horn. The thresholds were set for the normal level and a minimum acceptable level. The minimum level alarm or no alarm signal prompts maintenance action. The SLARM's™ history provides proof of proper operation. Alarm events are time-stamped and logged.

Is It Loud Enough? Tornado, fire, nuclear power plant alarms and sirens, as well as many other public safety and industrial warning devices can benefit from monitoring. Using the SLARM's™ standalone operation and the ODM microphone assembly make these remote installations feasible.

A Stinky Problem. A Medivac helicopter on its life-saving mission quickly approaches the hospital helipad and sets down. On the ground, the helicopter engines idle, prepared for a quick response to the next emergency.

The problem: the exhaust fumes from the engines drift upward toward the HVAC vents eight stories above. Specialized carbon filters and engineering staff run to the HVAC controls to turn them off—often forgetting to turn them back on, costing the hospital over $50,000 a year and hundreds of manhours.

A standalone SLARM™ with an ODM microphone mounted on the edge of the helipad detects arriving helicopters and turns off the HVAC intakes. As the helicopter departs, the vents are turned back on automatically. The SLARM™ not only provides control of the HVAC but also logs the arrival and departure events for future review, Fig. 2.9.

Too Much of a Good Thing Is a Problem. Noise complaints are often the result of *Too Much of a Good Thing.* A nightclub

FIGURE 2.9 ODM microphone assembly mounted on helipad. Courtesy ACO Pacific.

housed on the ground floor of a condo complex faced increased complaints from both condo owners and patrons alike.

The installation of a SLARM™ connected to the DJ's and sound staff's PC allowed them to monitor actual sound levels and alert them to excessive noise. The SLARM's™ positive indication of compliance assures maintenance of proper sound levels.

Protecting the Audience. Community and national regulations often specify noise limits for patrons and employees alike. Faced with the need to ensure that their audience's hearing was not damaged by *Too Much of a Good Thing*, a major broadcast company chose the SLARMSolution™.

Two SLARMS™ were used to monitor stage and auditorium levels. These units made use of both SPL and Leq Alarm settings. In addition, SLARMAnalysis™ is utilized to extrapolate daily Leq and dose estimates. The installations used the standard SLARM™ mic package and ACO Pacific's 7052PH phantom microphone system. The phantom system utilized the miles of microphone cables running through the complex. This made microphone placement easier. The results were proof of compliance, and the assurance that audience ears were not damaged.

NAMM 2008—Actual Measurements from the Show Floor. A SLARM™ was installed in a booth at the Winter NAMM 2008 show in Anaheim, CA. The microphone was placed at the back of the booth about 8 ft above the ground, away from the booth traffic and noise of people talking.

Figures 2.10–2.12 utilize SLARMWatch's™ History display capability as well as the SLARMAnalysis™ package. The SLARM™ operated standalone in the booth with the front panel LEDs advising the booth staff of critical noise levels.

The charts show the results of all four days of NAMM and Day 2. Day 2 was extracted from the data using the Zoom feature in SLARMWatch™. The booth was powered down in the evening, thus the quiet periods shown and the break in the history

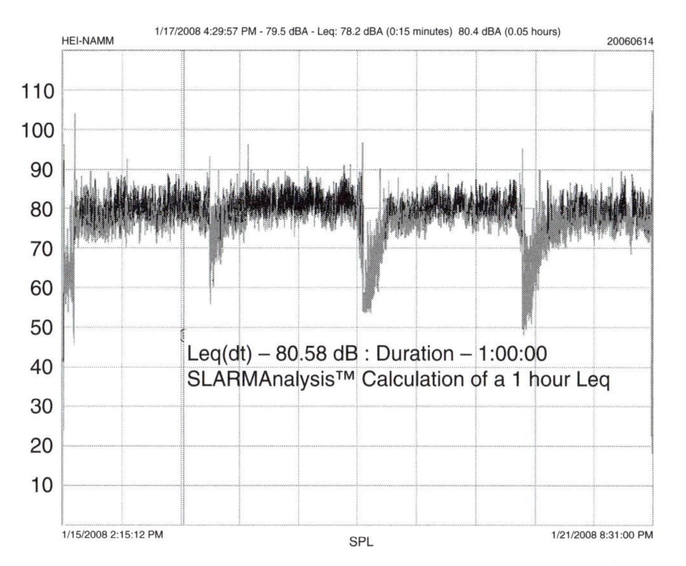

FIGURE 2.10 This is a dBA (A-weighted SPL) for all four days of NAMM. The booth power was shut down in the evening and then turned on for the exhibition. The SLARM™ restarted itself each morning and logged automatically during this time. It was not connected to a computer. The black indications are of sound levels exceeding the thresholds set in the SLARM™. Courtesy ACO Pacific.

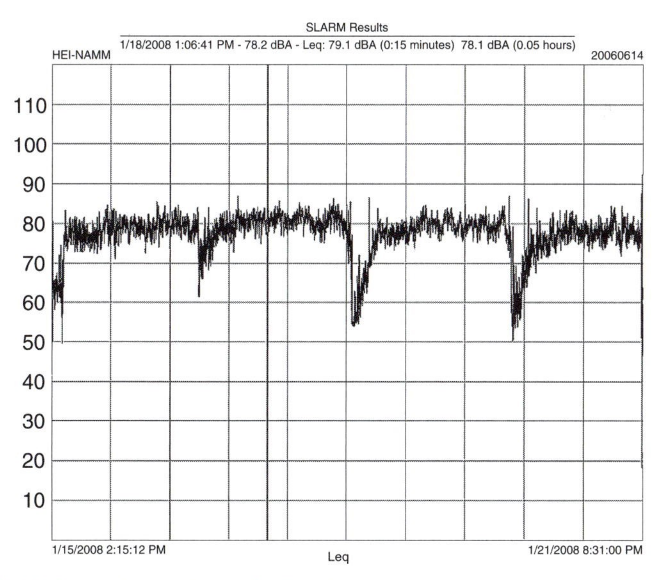

FIGURE 2.11 All four days 15 s LeqA. Courtesy ACO Pacific.

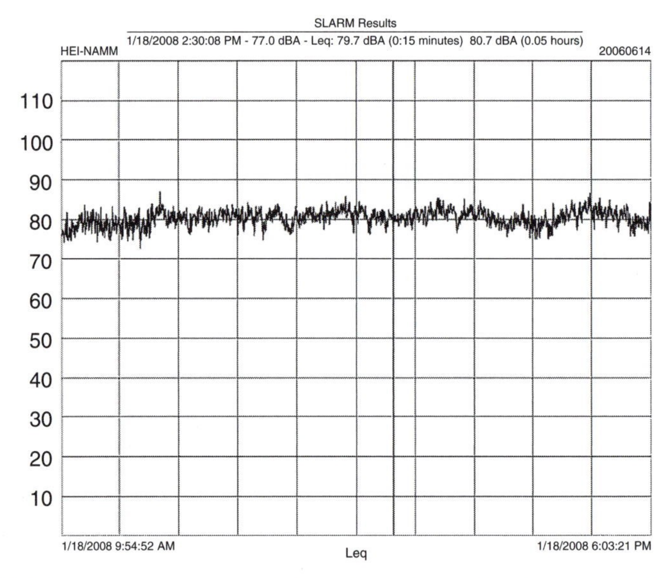

FIGURE 2.12 Day 2—a typical day. This chart is the Leq (15 s) dBA. This basically represents the running average sound level. Courtesy ACO Pacific.

sequence. The floor traffic quickly picked up at the beginning of the show day.

An 8 hour exposure at these levels has the potential to cause permanent hearing damage. The booth was located in one of the quieter areas of the NAMM Exhibition floor. Levels on the main show floor were at least 10–15 dB higher than those shown on the graphs.

2.6 SUMMARY

We live in a world of sounds and noise. Some enjoyable, some annoying, and all potentially harmful to health. Devices like the SLARM™ represent a unique approach to sound control and monitoring and are a useful tool for sound and noise pollution control. We hope we have provided insight into how much sound—noise to some—is part of our world to enjoy responsibly, While alerting you to the potential harm that sound can cause.

Fundamentals and Units of Measurement

Glen Ballou

3.1 UNITS OF MEASUREMENT

Measurements are the method we use to define all things in life. A dimension is any measurable extent such as length, thickness, or weight. A measurement system is any group of related unit names that state the quantity of properties for the items we see, taste, hear, smell, or touch.

A unit of measurement is the size of a quantity in the terms of which that quantity is measured or expressed, for instance, inches, miles, centimeters, and meters.

The laws of physics, which includes sound, are defined through dimensional equations that are defined from their units of measurements of mass, length, and time. For instance,

$$Area = L \times W$$

$$Velocity = \frac{D}{T}$$

where,

L is length,
W is width,
D is distance,
T is time.

A physical quantity is specified by a number and a unit: for instance, 16 ft or 5 m.

3.1.1 SI System

The SI system (from the French Système International d'Unités) is the accepted international modernized metric system of measurement. It is used worldwide with the exception of a few countries including the United States of America.

The SI system has the following advantages:

1. It is internationally accepted.
2. All values, except time, are decimal multiples or submultiples of the basic unit.
3. It is easy to use.
4. It is easy to teach.
5. It improves international trade and understanding.
6. It is coherent. All derived units are formed by multiplying and dividing other units without introducing any numerical conversion factor except one.
7. It is consistent. Each physical quantity has only one primary unit associated with it.

When using the SI system, exponents or symbol prefixes are commonly used. Table 3.1 is a chart of the accepted name of the number, its exponential form, symbol, and prefix name. (Note: because of their size, the numbers from sextillion to centillion have not been shown in numerical form and symbols and prefix names have not been established for these numbers.)

TABLE 3.1 Multiple and Submultiple Prefixes

Name of Number	Number	Exponential Form	Symbol	Prefix
Centillion		1.0×10^{303}		
Googol		1.0×10^{100}		
Vigintillion		1.0×10^{63}		
Novemdecillion		1.0×10^{60}		
Octodecillion		1.0×10^{57}		
Septendecillion		1.0×10^{54}		
Sexdecillion		1.0×10^{51}		
Quindecillion		1.0×10^{48}		
Quattuordecillion		1.0×10^{45}		
Tredecillion		1.0×10^{42}		
Duodecillion		1.0×10^{39}		
Undecillion		1.0×10^{36}		
Decillion		1.0×10^{33}		
Nonillion		1.0×10^{30}		
Octillion		1.0×10^{27}		
Septillion		1.0×10^{24}	E	Exa-
Sextillion		1.0×10^{21}	P	Peta-
Trillion	1,000,000,000,000	1.0×10^{12}	T	Tera-
Billion	1,000,000,000	1.0×10^{9}	G	Giga-
Million	1,000,000	1.0×10^{6}	M	Mega-
Thousand	1000	1.0×10^{3}	k	Kilo-
Hundred	100	1.0×10^{2}	h	Hecto-
Ten	10	1.0×10^{1}	da	Deka-
Unit	1	1.0×10^{0}	—	—
Tenth	0.10	1.0×10^{-1}	d	Deci-
Hundredth	0.01	1.0×10^{-2}	c	Centi-
Thousandth	0.001	1.0×10^{-3}	m	Milli-

TABLE 3.1 Multiple and Submultiple Prefixes—Contd.

Name of Number	Number	Exponential Form	Symbol	Prefix
Millionth	0.000 001	1.0×10^{-6}	μ	Micro-
Billionth	0.000 000 001	1.0×10^{-9}	n	Nano-
Trillionth	0.000 000 000 001	1.0×10^{-12}	p	Pico-
Quadrillionth	0.000 000 000 000 001	1.0×10^{-15}	f	Femto-

3.1.2 Fundamental Quantities

There are seven fundamental quantities in physics: length, mass, time, intensity of electric current, temperature, luminous intensity, and molecular substance. Two supplementary quantities are plane angle and solid angle.

3.1.3 Derived Quantities

Derived quantities are those defined in terms of the seven fundamental quantities, for instance, speed = length/time. There are sixteen derived quantities with names of their own: energy (work, quantity of heat), force, pressure, power, electric charge, electric potential difference (voltage), electric resistance, electric conductance, electric capacitance, electric inductance, frequency, magnetic flux, magnetic flux density, luminous flux, illuminance, and customary temperature. Following are thirteen additional derived quantities that carry the units of the original units that are combined. They are area, volume, density, velocity, acceleration, angular velocity, angular acceleration, kinematic viscosity, dynamic viscosity, electric field strength, magnetomotive force, magnetic field strength, and luminance.

3.1.4 Definition of the Quantities

The quantities will be defined in SI units, and their U.S. customary unit equivalent values will also be given.

Length (L). *Length* is the measure of how long something is from end to end. The meter (abbreviated m) is the SI unit of length. (Note: in the United States the spelling "meter" is retained, while most other countries use the spelling "metre.") The meter is the 1 650 763.73 wavelengths, in vacuum, of the radiation corresponding to the unperturbed transition between energy level $2P_{10}$ and $5D_5$ of the krypton-86 atom. The result is an orange-red line with a wavelength of 6057.802×10^{-10} meters. The meter is equivalent to 39.370 079 inches.

Mass (M). *Mass* is the measure of the inertia of a particle. The mass of a body is defined by the equation:

$$M = \left(\frac{A_s}{a}\right) M_s \qquad (3.1)$$

where,

A_s is the acceleration of the standard mass M_s,

a is the acceleration of the unknown mass, M, when the two bodies interact.

The kilogram (kg) is the unit of mass. This is the only base or derived unit in the SI system that contains a prefix. Multiples are formed by attaching prefixes to the word gram. Small masses may be described in grams (g) or milligrams (mg) and large masses in megagrams. Note the term *tonnes* is sometimes used for the metric ton or megagram, but this term is not recommended.

The present international definition of the kilogram is the mass of a special cylinder of platinum iridium alloy maintained at the International Bureau of Weights and Measures, Sevres, France. One kilogram is equal to 2.204 622 6 avoirdupois pounds (lb). A liter of pure water at standard temperature and pressure has a mass of 1 kg ± one part in 10^4.

Mass of a body is often revealed by its weight, which the gravitational attraction of the earth gives to that body.

If a mass is weighed on the moon, its mass would be the same as on earth, but its weight would be less due to the small amount of gravity.

$$M = \frac{W}{g} \tag{3.2}$$

where,

W is the weight,

g is the acceleration due to gravity.

Time (t). *Time* is the period between two events or the point or period during which something exists, happens, etc.

The second (s) is the unit of time. Time is the one dimension that does not have powers of ten multipliers in the SI system. Short periods of time can be described in milliseconds (ms) and microseconds (μs). Longer periods of time are expressed in minutes (1 min = 60 s) and hours (1 h = 3600 s). Still longer periods of time are the day, week, month, and year. The present international definition of the second is the time duration of 9, 192, 631, 770 periods of the radiation corresponding to the transition between the two hyperfine levels of the ground state of the atom of caesium 133. It is also defined as 1/86, 400 of the mean solar day.

Current (I). *Current* is the rate of flow of electrons. The ampere (A) is the unit of measure for current. Small currents are measured in milliamperes (mA) and microamperes (μA), and large currents are in kiloamperes (kA). The international definition of the ampere is the constant current that, if maintained in two straight parallel conductors of infinite length and negligible cross-sectional area and placed exactly 1 m apart in a vacuum, will produce between them a force of 2×10^{-7} N/m^2 of length.

A simple definition of one ampere of current is the intensity of current flow through a 1 ohm resistance under a pressure of 1 volt of potential difference.

Temperature (T). *Temperature* is the degree of hotness or coldness of anything. The kelvin (K) is the unit of temperature. The kelvin is 1/273.16 of the thermodynamic temperature of the triple point of pure water. Note: the term degree (°) is not used with the term kelvin as it is with other temperature scales.

Ordinary temperature measurements are made with the celsius scale on which water freezes at 0°C and boils at 100°C. A change of 1°C is equal to a change of 1 kelvin, therefore 0°C = 273.15 K: 0°C = 32°F.

Luminous Intensity (I_L). *Luminous intensity* is the luminous flux emitted per unit solid angle by a point source in a given direction. The candela (cd) is the unit of luminous intensity. One candela will produce a luminous flux of 1 lumen within a solid angle of 1 steradian.

The international definition of the candela is the luminous intensity, perpendicular to the surface, of 1/600 000 m^2 of a black body at the temperature of freezing platinum under a pressure of 101 325 N/m^2 (pascals).

Molecular Substance (n). *Molecular substance* is the amount of substance of a system that contains as many elementary entities as there are atoms in 0.012 kg of carbon 12.

The mole is the unit of molecular substance. One mole of any substance is the gram molecular weight of the material. For example, 1 mole of water (H_2O) weighs 18.016 g.

$$
\begin{aligned}
H_2 &= 2 \text{ atoms } \times 1.008 \text{ atomic weight} \\
O &= 2 \text{ atoms } \times 16 \text{ atomic weight} \\
H_2O &= 18.016\, g
\end{aligned}
$$

Plane Angle (α). The *plane angle* is formed between two straight lines or surfaces that meet. The radian (rad) is the unit of plane angles. One radian is the angle formed between two radii of a circle and subtended by an arc whose length is equal to the radius. There are 2π radians in 360°.

Ordinary measurements are still made in degrees. The degree can be divided into minutes and seconds or into tenths and hundredths of a degree. For small angles, the latter is most useful.

$$\text{One degree of arc } (1^\circ) = \frac{\pi}{180}\text{Rad}$$
$$1\,\text{Rad} = 57.2956^\circ \tag{3.3}$$

Solid Angle (A). A *solid angle* subtends three dimensions. The solid angle is measured by the area, subtended (by projection) on a sphere of unit radius by the ratio of the area A, intercepted on a sphere of radius r to the square of the radius (A/r^2).

The steradian (sr) is the unit of solid angle. The steradian is the solid angle at the center of a sphere that subtends an area on the spherical surface, which is equal to that of a square whose sides are equal to the radius of the sphere.

Energy (E). *Energy* is the property of a system that is a measure of its ability to do work. There are two main forms of energy—potential energy and kinetic energy.

1. Potential energy (U) is the energy possessed by a body or system by virtue of position and is equal to the work done in changing the system from some standard configuration to its present state. Potential energy is calculated with the equation:

$$U = Mgh \tag{3.4}$$

where:

M is the mass,
g is the acceleration due to gravity,
h is the height.

For example, a mass M placed at a height (h) above a datum level in a gravitational field with an acceleration of free fall (g), has a potential energy given by $U = mgh$. This potential energy is converted into kinetic energy when the body falls between the levels.

2. Kinetic energy (T) is the energy possessed by virtue of motion and is equal to the work that would be required to bring the body

to rest. A body undergoing translational motion with velocity, v, has a kinetic energy given by:

$$T = 0.5Mv^2 \tag{3.5}$$

where:

M is the mass of the body,
v is the velocity of the body.

For a body undergoing rotational motion:

$$T = 0.51I\omega^2 \tag{3.6}$$

where:

I is the moment of inertia of the body about its axis of rotation,
ω is the angular velocity.

The joule (J) is the unit of energy. The mechanical definition is the work done when the force of 1 newton is applied for a distance of 1 m in the direction of its application, or 1 Nm. The electrical unit of energy is the kilowatt-hour (kWh), which is equal to 3.6×10^6 J.

In physics, the unit of energy is the electron volt (eV), which is equal to $(1.602\,10 \pm 0.000\,07) \times 10^{-19}$ J.

Force (F). *Force* is any action that changes, or tends to change, a body's state of rest or uniform motion in a straight line.

The newton (N) is the unit of force and is that force which, when applied to a body having a mass of 1 kg, gives it an acceleration of 1 m/s^2. One newton equals 1 J/m, 1 kg(m)/s^2, 10^5 dynes, and 0.224 809 lb force.

Pressure. *Pressure* is the force (in a fluid) exerted per unit area on an infinitesimal plane situated at the point. In a fluid at rest, the pressure at any point is the same in all directions. A fluid is any material substance which in static equilibrium cannot exert tangential force across a surface but can exert only pressure. Liquids and gases are fluids.

The pascal (*Pa*) is the unit of pressure. The pascal is equal to the newton per square meter (N/m^2).

$$\begin{aligned} 1Pa &= 10^{-6}\text{bars} \\ &= 1.45038 \times 10^{-4}\text{lb/in}^2 \end{aligned} \tag{3.7}$$

Power (W). *Power* is the rate at which energy is expended or work is done. The watt (W) is the unit of power and is the power that generates energy at the rate of 1 J/s.

$$\begin{aligned} 1\text{W} &= 1\text{ J/s} \\ &= 3.414\,42\text{ BTU/h} \\ &= 44.2537\text{ft-lb/min} \\ &= 0.00134102\text{ hp} \end{aligned} \tag{3.8}$$

Electric Charge (Q). *Electric charge* is the quantity of electricity or electrons that flows past a point in a period of time. The coulomb (*C*) is the unit of electric charge and is the quantity of electricity moved in 1 second by a current of 1 ampere. The coulomb is also defined as 6.24196×10^{18} electronic charges.

Electric Potential Difference (V). Often called electromotive force (*emf*) and voltage (*V*), *electric potential difference* is the line integral of the electric field strength between two points. The volt (*V*) is the unit of electric potential. The volt is the potential difference that will cause a current flow of 1 A between two points in a circuit when the power dissipated between those two points is 1 W.

A simpler definition would be to say a potential difference of 1 V will drive a current of 1 A through a resistance of 1 Ω.

$$\begin{aligned} Volt(V) &= \frac{W}{A} \\ &= \frac{J}{A(s)} \\ &= \frac{kg\,(m^2)}{s^3 A} \\ &= A\Omega \end{aligned} \tag{3.9}$$

Electric Resistance (R). *Electric resistance* is the property of conductors that, depending in their dimensions, material, and temperature, determines the current produced by a given difference of potential. It is also that property of a substance that impedes current and results in the dissipation of power in the form of heat.

The ohm (Ω) is the unit of resistance and is the resistance that will limit the current flow to 1 A when a potential difference of 1 V is applied to it.

$$\begin{aligned} R &= \frac{V}{A} \\ &= \frac{kg\,(m^2)}{s^3 A^3} \end{aligned} \tag{3.10}$$

Electric Conductance (G). *Electric conductance* is the reciprocal of resistance. The siemens (*S*) is the unit of electric conductance. A passive device that has a conductance of 1 S will allow a current flow of 1 A when 1 V potential is applied to it.

$$\begin{aligned} S &= \frac{1}{\Omega} \\ &= \frac{A}{V} \end{aligned} \tag{3.11}$$

Electric Capacitance (C). *Electric capacitance* is the property of an isolated conductor or set of conductors and insulators to store electric charge. The farad (*F*) is the unit of electric capacitance and is defined as the capacitance that exhibits a potential difference of 1 V when it holds a charge of 1 C.

$$\begin{aligned} F &= \frac{C}{V} \\ &= \frac{AS}{V} \end{aligned} \tag{3.12}$$

where:

C is the electric charge in coulombs,

V is the electric potential difference in volts,

A is the current in amperes,
S is the conductance in siemens.

Electric Inductance (L). *Electric inductance* is the property that opposes any change in the existing current. Inductance is only present when the current is changing. The henry (*H*) is the unit of inductance and is the inductance of a circuit in which an electromotive force of 1 V is developed by a current change of 1 A/s.

$$H = \frac{Vs}{A} \tag{3.13}$$

Frequency (f). *Frequency* is the number of recurrences of a periodic phenomenon in a unit of time. The hertz (*Hz*) is the unit of frequency and is equal to one cycle per second, 1 Hz = 1 cps. Frequency is often measured in hertz (Hz), kilohertz (kHz), and megahertz (MHz).

Sound Intensity (W/m²). *Sound intensity* is the rate of flow of sound energy through a unit area normal to the direction of flow. For a sinusoidally varying sound wave the intensity I is related to the sound pressure p and the density β of the medium by

$$I = \frac{p^2}{\beta c} \tag{3.14}$$

where:

c is the velocity of sound.

The watt per square meter (W/m^2) is the unit of sound intensity.

Magnetic Flux (ϕ). *Magnetic flux* is a measure of the total size of a magnetic field. The weber (*Wb*) is the unit of magnetic flux, and is the amount of flux that produces an electromotive force of 1 V in a one-turn conductor as it reduces uniformly to zero in 1 s.

$$\begin{aligned} Wb &= W(s) \\ &= 10^8 \text{ lines of flux} \\ &= \frac{kg(m^2)}{s^2 A} \end{aligned} \tag{3.15}$$

Magnetic Flux Density (ß). The *magnetic flux density* is the flux passing through the unit area of a magnetic field in the direction at right angles to the magnetic force. The vector product of the magnetic flux density and the current in a conductor gives the force per unit length of the conductor.

The tesla (T) is the unit of magnetic flux density and is defined as a density of 1 Wb/m^2.

$$\begin{aligned} T &= \frac{Wb}{m^2} \\ &= \frac{V(s)}{m^2} \\ &= \frac{kg}{s^2 A} \end{aligned} \tag{3.16}$$

Luminous Flux (Φ_v). *Luminous flux* is the rate of flow of radiant energy as evaluated by the luminous sensation that it produces. The lumen (lm) is the unit of luminous flux, which is the amount of luminous flux emitted by a uniform point source whose intensity is 1 steradian.

$$\begin{aligned} lm &= cd\left(\frac{sr}{m^2}\right) \\ &= 0.0795774 \text{ candlepower} \end{aligned} \tag{3.17}$$

where:

cd is the luminous intensity in candelas,
sr is the solid angle in steradians.

Luminous Flux Density (E_v). The *luminous flux density* is the luminous flux incident on a given surface per unit area. It is sometimes called illumination or intensity of illumination. At any point on a surface, the illumination is given by:

$$E_v = \frac{d\Phi_v}{dA} \tag{3.18}$$

The lux (lx) is the unit of luminous flux density, which is the density of radiant flux of lm/m^2,

$$
\begin{aligned}
lx &= \frac{lm}{m^2} \\
&= cd\frac{sr}{m^2} \qquad (3.19) \\
&= 0.0929030fc
\end{aligned}
$$

Displacement. *Displacement* is a change in position or the distance moved by a given particle of a system from its position of rest, when acted on by a disturbing force.

Speed/Velocity. *Speed* is the rate of increase of distance traveling by a body. Average speed is found by the equation:

$$S = \frac{l}{t} \qquad (3.20)$$

where:

S is the speed,
l is the length or distance,
t is the time to travel.

Speed is a scalar quantity as it is not referenced to direction. Instantaneous speed = dl/dt. *Velocity* is the rate of increase of distance traversed by a body in a particular direction.

Velocity is a vector quantity as both speed and direction are indicated. The l/t can often be the same for the velocity and speed of an object. However, when speed is given, the direction of movement is not known. If a body describes a circular path and each successive equal distance along the path is described in equal times, the speed would be constant but the velocity would constantly change due to the change in direction.

Weight. *Weight* is the force exerted on a mass by the gravitational pull of the planet, star, moon, etc., that the mass is near. The weight experienced on earth is due to the earth's gravitational pull, which is 9.806 65 m/s^2, and causes an object to accelerate toward earth at a rate of 9.806 65 m/s^2 or 32 ft/s^2.

The weight of a mass M is M(g). If M is in kg and g in m/s^2, the weight would be in newtons (N). Weight in the U.S. system is in pounds (lb).

Acceleration. *Acceleration* is the rate of change in velocity or the rate of increase or decrease in velocity with time. Acceleration is expressed in meters per second squared (m/s^2), or ft/s^2 in the U.S. system.

Amplitude. *Amplitude* is the magnitude of variation in a changing quantity from its zero value. Amplitude should always be modified with adjectives such as peak, rms, maximum, instantaneous, etc.

Wavelength (M). In a periodic wave, the distance between two points of the corresponding phase of two consecutive cycles is the *wavelength.* Wavelength is related to the velocity of propagation (c) and frequency (f) by the equation:

$$\lambda = \frac{c}{f} \tag{3.21}$$

The wavelength of a wave traveling in air at sea level and standard temperature and pressure (STP) is:

$$\lambda = \frac{331.4\,\text{m/s}}{f} \tag{3.22}$$

or

$$\lambda = \frac{1087.42\,\text{ft/s}}{f} \tag{3.23}$$

For instance, the length of a 1000 Hz wave would be 0.33 m, or 1.09 ft.

Phase. *Phase* is the fraction of the whole period that has elapsed, measured from a fixed datum. A sinusoidal quantity may be expressed as a rotating vector *OA*. When rotated a full 360 degrees, it represents a sine wave. At any position around the circle, *OX* is equal in length but said to be *X* degrees out of phase with *OA*.

It may also be stated that the phase difference between *OA* and *OX* is α. When particles in periodic motion due to the passage of a wave are moving in the same direction with the same relative displacement, they are said to be in phase. Particles in a wave front are in the same phase of vibration when the distance between consecutive wave fronts is equal to the wavelength. The phase difference of two particles at distances X_1 and X_2 is:

$$\alpha = \frac{2\pi(X_2 - X_1)}{\lambda} \tag{3.24}$$

Periodic waves, having the same frequency and waveform, are said to be in phase if they reach corresponding amplitudes simultaneously.

Phase Angle. The angle between two vectors representing two periodic functions that have the same frequency is the *phase angle*. Phase angle can also be considered the difference, in degrees, between corresponding stages of the progress of two cycle operations.

Phase Difference (ϕ). *Phase difference* is the difference in electrical degrees or time, between two waves having the same frequency and referenced to the same point in time.

Phase Shift. Any change that occurs in the phase of one quantity or in the phase difference between two or more quantities is the *phase shift*.

Phase Velocity. The *phase velocity* is when a point of constant phase is propagated in a progressive sinusoidal wave.

Temperature. *Temperature* is the measure of the amount of coldness or hotness. While kelvin is the SI standard, temperature is commonly referenced as °C (degrees Celsius) or °F (degrees Fahrenheit).

The lower fixed point (the ice point) is the temperature of a mixture of pure ice and water exposed to the air at standard atmospheric pressure.

The upper fixed point (the steam point) is the temperature of steam from pure water boiling at standard atmospheric pressure.

In the Celsius scale, named after Anders Celsius (1701–1744) and originally called Centigrade, the fixed points are 0°C and 100°C. This scale is used in the SI system.

The Fahrenheit scale, named after Gabriel Daniel Fahrenheit in 1714, has the fixed points at 32°F and 212°F.

To interchange between °C and °F, use the following equations:

$$
\begin{aligned}
{}^\circ\mathrm{C} &= ({}^\circ\mathrm{F} - 32^\circ) \times \frac{5}{9} \\
{}^\circ\mathrm{F} &= \left({}^\circ\mathrm{C} \times \frac{9}{5}\right) + 32^\circ
\end{aligned}
\qquad (3.25)
$$

The absolute temperature scale operates from absolute zero of temperature. Absolute zero is the point where a body cannot be further cooled because all the available thermal energy is extracted.

Absolute zero is 0 kelvin (0 K) or 0° Rankine (0°R). The Kelvin scale, named after Lord Kelvin (1850), is the standard in the SI system and is related to °C.

$$0^\circ\mathrm{C} = 273.15\ \mathrm{K}$$

The Rankine scale is related to the Fahrenheit system.

$$32^\circ\mathrm{F} = 459.67^\circ\mathrm{R}$$

The velocity of sound is affected by temperature. As the temperature increases, the velocity increases. The approximate formula is

$$C = 331.4\ \mathrm{m/s} + 0.607\,T\ \mathrm{SI\,units} \qquad (3.26)$$

where:

T is the temperature in °C.

$$C = 1052 \text{ ft/s} \times 1.106T \text{ U.S units} \quad (3.27)$$

where:

T is the temperature in °F.

Another simpler equation to determine the velocity of sound is:

$$C = 49.00\sqrt{459.69° + °F} \quad (3.28)$$

Things that can affect the speed of sound are the sound wave going through a temperature barrier or going through a stream of air such as from an air conditioner. In either case, the wave is deflected the same way that light is refracted in glass.

Pressure and altitude do not affect the speed of sound because at sea level the molecules bombard each other, slowing down their speed. At upper altitudes they are farther apart so they do not bombard each other as often, so they reach their destination at the same time.

Thevenin's Theorem. *Thevenin's Theorem* is a method used for reducing complicated networks to a simple circuit consisting of a voltage source and a series impedance. The theorem is applicable to both ac and dc circuits under steady-state conditions.

The theorem states: the current in a terminating impedance connected to any network is the same as if the network were replaced by a generator with a voltage equal to the open-circuit voltage of the network, and whose impedance is the impedance seen by the termination looking back into the network. All generators in the network are replaced with impedance equal to the internal impedances of the generators.

Kirchhoff's Laws. The laws of Kirchhoff can be used for both dc and ac circuits. When used in ac analysis, phase must also be taken into consideration.

Kirchhoff's Voltage Law (KVL). *Kirchhoff's voltage law* states that the sum of the branch voltages for any closed loop is zero at any time. Stated another way, for any closed loop, the sum of the voltage drops equal the sum of the voltage rises at any time.

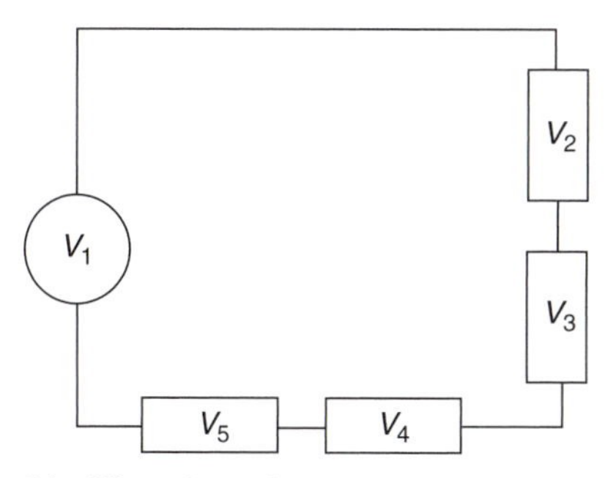

FIGURE 3.1 Kirchhoff's voltage law.

In the laws of Kirchhoff, individual electric circuit elements are connected according to some wiring plan or schematic. In any closed loop, the voltage drops must be equal to the voltage rises. For example, in the dc circuit of Fig. 3.1, V_1 is the voltage source or rise such as a battery and V_2, V_3, V_4, and V_5 are voltage drops (possibly across resistors) so:

$$V_1 = V_2 + V_3 + V_4 + V_5 \tag{3.29}$$

or:

$$V_1 - V_2 - V_3 - V_4 - V_5 = 0 \tag{3.30}$$

In an ac circuit, phase must be taken into consideration, therefore, the voltage would be:

$$V_1 e^{j\omega t} - V_2 e^{j\omega t} - V_3 e^{j\omega t} - V_4 e^{j\omega t} - V_5 e^{j\omega t} = 0 \tag{3.31}$$

where:

$e^{j\omega t}$ is $\cos At + j\sin At$ or Euler's identity.

Kirchhoff's Current Law (KCL). *Kirchhoff's current law* states that the sum of the branch currents leaving any node must equal the sum of the branch currents entering that node at any time.

Stated another way, the sum of all branch currents incident at any node is zero.

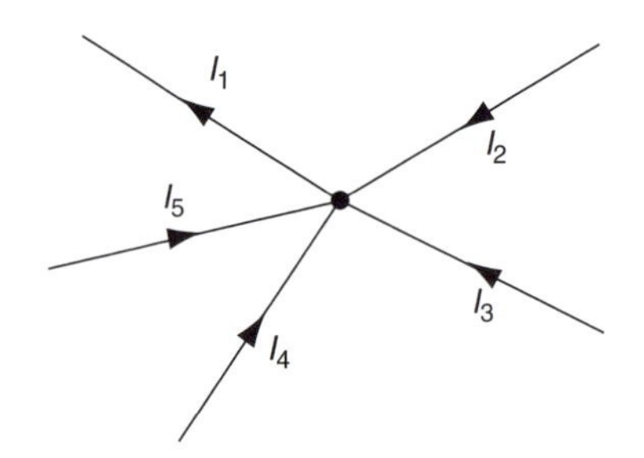

FIGURE 3.2 Kirchhoff's current law.

In Fig. 3.2 the connection on node current in a dc circuit is equal to 0 and is equal to the sum of currents I_1, I_2, I_3, I_4, and I_5 or:

$$I_1 = I_2 + I_3 + I_4 + I_5 \tag{3.32}$$

or:

$$I_1 - I_2 - I_3 - I_4 - I_5 = 0 \tag{3.33}$$

The current throughout the circuit is also a function of the current from the power source (V_1) and the current through all of the branch circuits.

In an ac circuit, phase must be taken into consideration, therefore, the current would be:

$$I_1 e^{j\omega t} - I_2 e^{j\omega t} - I_3 e^{j\omega t} - I_4 e^{j\omega t} - I_5 e^{j\omega t} = 0 \tag{3.34}$$

where:

$e^{j\omega t}$ is $\cos At + j\sin At$ or Euler's identity.

Ohm's Law. *Ohm's Law* states that the ratio of applied voltage to the resultant current is a constant at every instant and that this ratio is defined to be the resistance.

If the voltage is expressed in volts and the current in amperes, the resistance is expressed in ohms. In equation form it is:

$$R = \frac{V}{I} \tag{3.35}$$

or:

$$R = \frac{e}{i} \tag{3.36}$$

where:

e and i are instantaneous voltage and current,
V and I are constant voltage and current,
R is the resistance.

Through the use of Ohm's Law, the relationship between voltage, current, resistance or impedance, and power can be calculated.

Power is the rate of doing work and can be expressed in terms of potential difference between two points (voltage) and the rate of flow required to transform the potential energy from one point to the other (current). If the voltage is in volts or J/C and the current is in amperes or C/s, the product is joules per second or watts:

$$P = VI \tag{3.37}$$

or:

$$\frac{J}{s} = \frac{J}{C}\left(\frac{C}{s}\right) \tag{3.38}$$

where:

J is energy in joules,
C is electric charge in coulombs.

Fig. 3.3 is a wheel chart that relates current, voltage, resistance or impedance, and power. The power factor (PF) is cos I where I is the phase angle between e and i. A power factor is required in ac circuits.

3.2 RADIO FREQUENCY SPECTRUM

The *radio frequency spectrum* of 30 Hz– 3,000,000 MHz (3×10^{12} Hz) is divided into the various bands shown in Table 3.2.

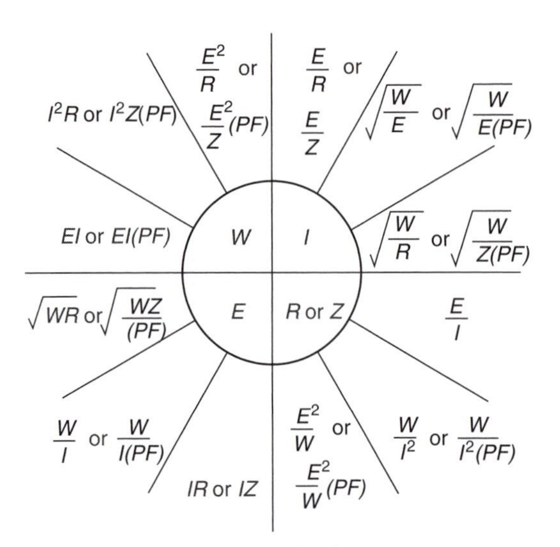

FIGURE 3.3 Power, voltage, current wheel.

TABLE 3.2 Frequency Classification

Frequency	Band No.	Classification	Abbreviation
30–300 Hz	2	extremely low frequencies	ELF
300–3000 Hz	3	voice frequencies	VF
3–30 kHz	4	very low frequencies	VLF
30–300 kHz	5	low frequencies	LF
300–3000 kHz	6	medium frequencies	MF
3–30 MHz	7	high frequencies	HF
30–300 MHz	8	very high frequencies	VHF
30–3000 MHz	9	ultrahigh frequencies	UHF
3–30 GHz	10	super-high frequencies	SHF
30–300 GHz	11	extremely high frequencies	EHF
300–3 THz	12	–	–

3.3 DECIBEL (dB)

Decibels are a logarithmic ratio of two numbers. The decibel is derived from two power levels and is also used to show voltage ratios indirectly (by relating voltage to power). The equations or decibels are

$$Power\ dB = 10\log\frac{P_1}{P_2} \tag{3.39}$$

$$Voltage\ dB_v = 20\log\frac{E_1}{E_2} \tag{3.40}$$

Fig. 3.4 shows the relationship between the power, decibels, and voltage. In the illustration, "dBm" is the decibels referenced to 1 mW.

Table 3.3 shows the relationship between decibel, current, voltage, and power ratios.

Volume unit (VU) meters measure decibels that are related to a 600 Ω impedance, O VU is actually +4 dBm. When measuring decibels referenced to 1 mW at any other impedance than 600 Ω, use:

$$dBm\ at\ new\ Z = dBm_{600\Omega} + 10\log\frac{600\,\Omega}{Z_{new}} \tag{3.41}$$

Example: The dBm for a 32 Ω load is:

$$\begin{aligned} dBm_{32} &= 4\ \text{dBm} + 10\log\frac{600\,\Omega}{32\,\Omega} \\ &= 16.75\,\text{dBm} \end{aligned}$$

This can also be determined by using the graph in Fig. 3.5.

To find the logarithm of a number to some other base than the base 10 and 2.718, use:

$$n = b^L \tag{3.42}$$

A number is equal to a base raised to its logarithm:

$$\ln(n) = \ln(bL) \tag{3.43}$$

therefore:

$$\frac{\ln(n)}{\ln(b)} = L \tag{3.44}$$

The natural log is a number divided by the natural log of the base equals the logarithm.

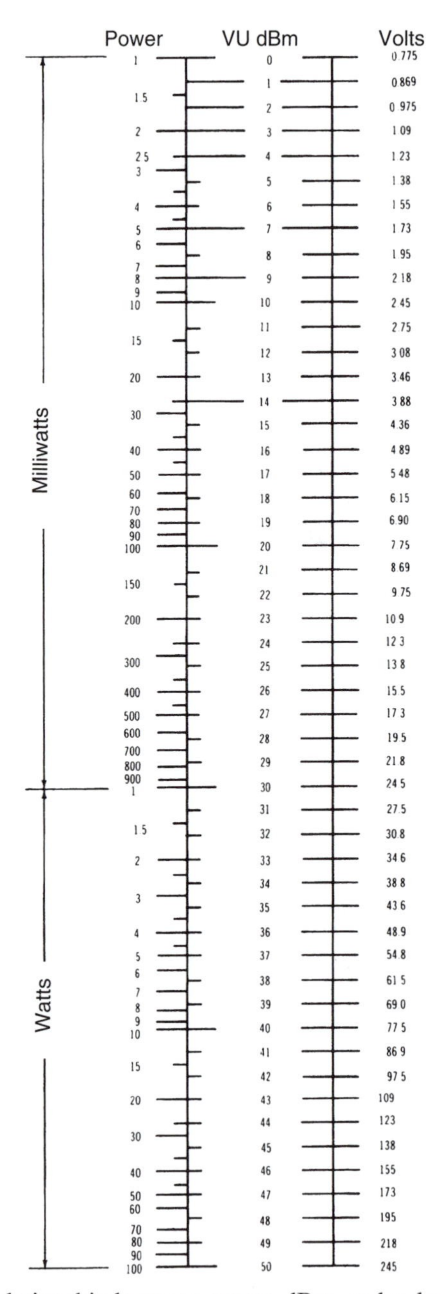

FIGURE 3.4 Relationship between power, dBm, and voltage.

TABLE 3.3 Relationships between Decibel, Current, Voltage, and Power Ratios

dB Voltage	Loss	Gain	dB Power	dB Voltage	Loss	Gain	dB Power	dB Voltage	Loss	Gain	dB Power	dB Voltage	Loss	Gain	dB Power
0.0	**1.0000**	**1.000**	**0.0**	**5.0**	**0.5623**	**1.778**	**0.50**	**10.0**	**0.3162**	**3.162**	**5.00**	**15.0**	**0.1778**	**5.623**	**0.50**
0.1	0.9886	1.012	0.05	0.1	0.5559	1.799	0.55	0.1	0.3126	3.199	0.05	0.1	0.1758	5.689	0.55
0.2	0.9772	1.023	0.10	0.2	0.5495	1.820	0.60	0.2	0.3090	3.236	0.10	0.2	0.1738	5.754	0.60
0.3	0.9661	1.035	0.15	0.3	0.5433	1.841	0.65	0.3	0.3055	3.273	0.15	0.3	0.1718	5.821	0.65
0.4	0.9550	1.047	0.20	0.4	0.5370	1.862	0.70	0.4	0.3020	3.311	0.20	0.4	0.1698	5.888	0.70
0.5	0.9441	1.059	0.25	0.5	0.5309	1.884	0.75	0.5	0.2985	3.350	0.25	0.5	0.1679	5.957	0.75
0.6	0.9333	1.072	0.30	0.6	0.5248	1.905	0.80	0.6	0.2951	3.388	0.30	0.6	0.1660	6.026	0.80
0.7	0.9226	1.084	0.35	0.7	0.5188	1.928	0.85	0.7	0.2917	3.428	0.35	0.7	0.1641	6.095	0.85
0.8	0.9120	1.096	0.40	0.8	0.5129	1.950	0.90	0.8	0.2884	3.467	0.40	0.8	0.1622	6.166	0.90
0.9	0.9016	1.109	0.45	0.9	0.5070	1.972	0.95	0.9	0.2851	3.508	0.45	0.9	0.1603	6.237	0.95
1.0	**0.8913**	**1.122**	**0.50**	**6.0**	**0.5012**	**1.995**	**3.00**	**11.0**	**0.2818**	**3.548**	**0.50**	**16.0**	**0.1585**	**6.310**	**8.00**
0.1	0.8810	1.135	0.55	0.1	0.4955	2.018	0.05	01	0.2786	3.589	0.55	01	0.1567	6.383	0.05
0.2	0.8710	1.148	0.60	0.2	0.4898	2.042	0.10	02	0.2754	3.631	0.60	02	0.1549	6.457	0.10
0.3	0.8610	1.161	0.65	0.3	0.4842	2.065	0.15	03	0.2723	3.673	0.65	03	0.1531	6.531	0.15
0.4	0.8511	1.175	0.70	0.4	0.4786	2.089	0.20	04	0.2692	3.715	0.70	04	0.1514	6.607	0.20

TABLE 3.3 Relationships between Decibel, Current, Voltage, and Power Ratios—Contd.

dB Voltage	Loss	Gain	dB Power	dB Voltage	Loss	Gain	dB Power	dB Voltage	Loss	Gain	dB Power	dB Voltage	Loss	Gain	dB Power
0.5	0.8414	1.189	0.75	0.5	0.4732	2.113	0.25	05	0.2661	3.758	0.75	05	0.1496	6.683	0.25
0.6	0.8318	1.202	0.80	0.6	0.4677	2.138	0.30	06	0.2630	3.802	0.80	06	0.1479	6.761	0.30
0.7	0.8222	1.216	0.85	0.7	0.4624	2.163	0.35	07	0.2600	3.846	0.85	07	0.1462	6.839	0.35
0.8	0.8128	1.230	0.90	0.8	0.4571	2.188	0.40	08	0.2570	3.890	0.90	08	0.1445	6.918	0.40
0.9	0.8035	1.245	0.95	0.9	0.4519	2.213	0.45	09	0.2541	3.936	0.95	09	0.1429	6.998	0.45
2.0	**0.7943**	**1.259**	**1.00**	**7.0**	**0.4467**	**2.239**	**0.50**	**12.0**	**0.2512**	**3.981**	**6.00**	**17.0**	**0.1413**	**7.079**	**0.50**
0.1	0.7852	1.274	0.05	0.1	0.4416	2.265	0.55	0.1	0.2483	4.027	0.05	0.1	0.1396	7.161	0.55
0.2	0.7762	1.288	0.10	0.2	0.4365	2.291	0.60	0.2	0.2455	4.074	0.10	0.2	0.1380	7.244	0.60
0.3	0.7674	1.303	0.15	0.3	0.4315	2.317	0.65	0.3	0.2427	4.121	0.15	0.3	0.1365	7.328	0.65
0.4	0.7586	1.318	0.20	0.4	0.4266	2.344	0.70	0.4	0.2399	4.169	0.20	0.4	0.1349	7.413	0.70
0.5	0.7499	1.334	0.25	0.5	0.4217	2.371	0.75	0.5	0.2371	4.217	0.25	0.5	0.1334	7.499	0.75
0.6	0.7413	1.349	0.30	0.6	0.4169	2.399	0.80	0.6	0.2344	4.266	0.30	0.6	0.1318	7.586	0.80
0.7	0.7328	1.365	0.35	0.7	0.4121	2.427	0.85	0.7	0.2317	4.315	0.35	0.7	0.1303	7.674	0.85
0.8	0.7244	1.380	0.40	0.8	0.4074	2.455	0.90	0.8	0.2291	4.365	0.40	0.8	0.1288	7.762	0.90
0.9	0.7161	1.396	0.45	0.9	0.4027	2.483	0.95	0.9	0.2265	4.416	0.45	0.9	0.1274	7.852	0.95

TABLE 3.3 Relationships between Decibel, Current, Voltage, and Power Ratios—Contd.

dB Voltage	Loss	Gain	dB Power	dB Voltage	Loss	Gain	dB Power	dB Voltage	Loss	Gain	dB Power	dB Voltage	Loss	Gain	dB Power
3.0	**0.7079**	**1.413**	**0.50**	**8.0**	**0.3981**	**2.512**	**4.00**	**13.0**	**0.2239**	**4.467**	**0.50**	**18.0**	**0.1259**	**7.943**	**9.00**
0.1	0.6998	1.429	0.55	0.1	0.3936	2.541	0.05	0.1	0.2213	4.519	0.55	0.1	0.1245	8.035	0.05
0.2	0.6918	1.445	0.60	0.2	0.3890	2.570	0.10	0.2	0.2188	4.571	0.60	0.2	0.1230	8.128	0.10
0.3	0.6839	1.462	0.65	0.3	0.3846	2.600	0.15	0.3	0.2163	4.624	0.65	0.3	0.1216	8.222	0.15
0.4	0.6761	1.479	0.70	0.4	0.3802	2.630	0.20	0.4	0.2138	4.677	0.70	0.4	0.1202	8.318	0.20
0.5	0.6683	1.496	0.75	0.5	0.3758	2.661	0.25	0.5	0.2113	4.732	0.75	0.5	0.1189	8.414	0.25
0.6	0.6607	1.514	0.80	0.6	0.3715	2.692	0.30	0.6	0.2089	4.786	0.80	0.6	0.1175	8.511	0.30
0.7	0.6531	1.531	0.85	0.7	0.3673	2.723	0.35	0.7	0.2065	4.842	0.85	0.7	0.1161	8.610	0.35
0.8	0.6457	1.549	0.90	0.8	0.3631	2.754	0.40	0.8	0.2042	4.898	0.90	0.8	0.1148	8.710	0.40
0.9	0.6383	1.567	0.95	0.9	0.3589	2.786	0.45	0.9	0.2018	4.955	0.95	0.9	0.1135	8.810	0.45
4.0	**0.6310**	**1.585**	**2.00**	**9.0**	**0.3548**	**2.818**	**0.50**	**14.0**	**0.1995**	**5.012**	**7.00**	**19.0**	**0.1122**	**8.913**	**0.50**
0.1	0.6237	1.603	0.05	0.1	0.3508	2.851	0.55	0.1	0.1972	5.070	0.05	0.1	0.1109	9.016	0.55
0.2	0.6166	1.622	0.10	0.2	0.3467	2.884	0.60	0.2	0.1950	5.129	0.10	0.2	0.1096	9.120	0.60
0.3	0.6095	1.641	0.15	0.3	0.3428	2.917	0.65	0.3	0.1928	5.188	0.15	0.3	0.1084	9.226	0.65
0.4	0.6026	1.660	0.20	0.4	0.3388	2.951	0.70	0.4	0.1905	5.248	0.20	0.4	0.1072	9.333	0.70

TABLE 3.3 Relationships between Decibel, Current, Voltage, and Power Ratios—Contd.

dB Voltage	Loss	Gain	dB Power	dB Voltage	Loss	Gain	dB Power	dB Voltage	Loss	Gain	dB Power	dB Voltage	Loss	Gain	dB Power
0.5	0. 5957	1.679	0.25	0.5	0.3350	2.985	0.75	0.5	0.1884	5.309	0.25	0.5	0.1059	9.441	0.75
0.6	0.5888	1.698	0.30	0.6	0.3311	3.020	0.80	0.6	0.1862	5.370	0.30	0.6	0.1047	9.550	0.80
0.7	0. 5821	1.718	0.35	0.7	0.3273	3.055	0.85	0.7	0.1841	5.433	0.35	0.7	0.1035	9.661	0.85
0.8	0. 5754	1.738	0.40	0.8	0.3236	3.090	0.90	0.8	0.1820	5.495	0.40	0.8	0.1023	9.772	0.90
0.9	0.5689	1.758	0.45	0.9	0.3199	3.126	0.95	0.9	0.1799	5.559	0.45	0.9	0.1012	9.886	0.95

dB Voltage	Loss	Gain	dB Power	dB Voltage	Loss	Gain	dB Power
20.0	**0.1000**	**10.00**	**10.00**	**60.0**	**0.001**	**1,000**	**30.00**
	Use the same number as 0–20 dB but shift decimal point one step to the left. Thus since	Use the same number as 0–20 dB but shift decimal point one step to the right. Thus since	This column repeats every 10 dB instead of 20 dB		Use the same numbers as 0–20 dB but shift point three steps to the left. Thus since	Use the same number as 0–20 dB column but shift point three steps to the right. Thus since	This column repeats every 10 dB instead of 20 dB
	10 dB = 0.3162	10 dB = 3.162			10 dB = 0.3162	10 dB = 3.162	
	30 dB = 0.03162	30 dB = 31.62			70 dB = 0.0003162	70 dB = 3162	

TABLE 3.3 Relationships between Decibel, Current, Voltage, and Power Ratios—Contd.

dB Voltage	Loss	Gain	dB Power	dB Voltage	Loss	Gain	dB Power
40.0	**0.01**	**100**	**20**	**80**	**0.0001**	**10,000**	**40.00**
	Use the same number as 0–20 dB but shift point two steps to the left. Thus since	Use the same number as 0–20 dB but shift point two steps to the right. Thus since	This column repeats every 10 dB instead of 20 dB		Use the same numbers as 0–20 dB but shift point four steps to the left. Thus since	Use the same number as 0–20 dB but shift point four steps to the right. Thus since	This column repeats every 10 dB instead of 20 dB
	10 dB = 03162	10 dB = 3162			10 dB = 0.3162	10 dB = 3.162	
	50 dB = 0.003162	50 dB = 316.2			90 dB = 0.00003162	90 dB = 31620	
				100	**0.00001**	**100,000**	**50.00**

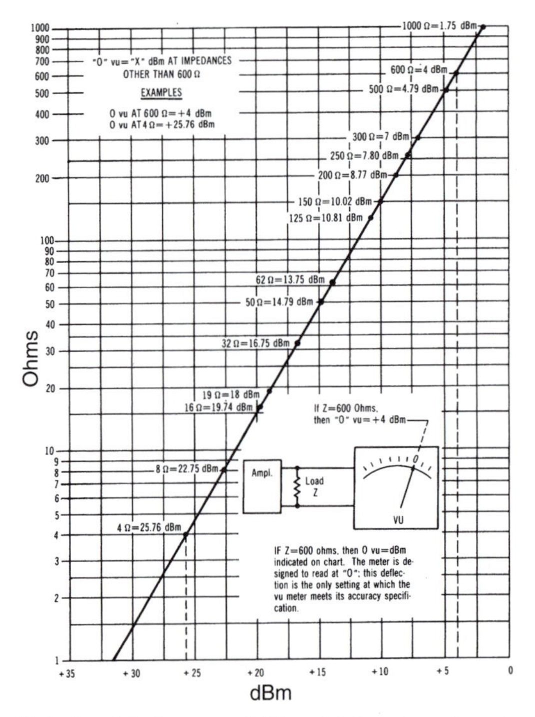

FIGURE 3.5 Relationship between VU and dBm at various impedances.

Example: Find the logarithm of the number 2 to the base 10:

$$\frac{\ln 2}{\ln 10} = \frac{0.693147}{2.302585}$$
$$= 0.301030$$

In information theory work, logarithms to the base 2 are quite commonly employed. To find the $\log_2$ of 26:

$$\frac{\ln 26}{\ln 2} = 4.70$$

To prove this, raise 2 to the 4.70 power:

$$2^{4.70} = 26$$

3.4 SOUND PRESSURE LEVEL

The *sound pressure level* (*SPL*) is related to acoustic pressure as seen in Fig. 3.6.

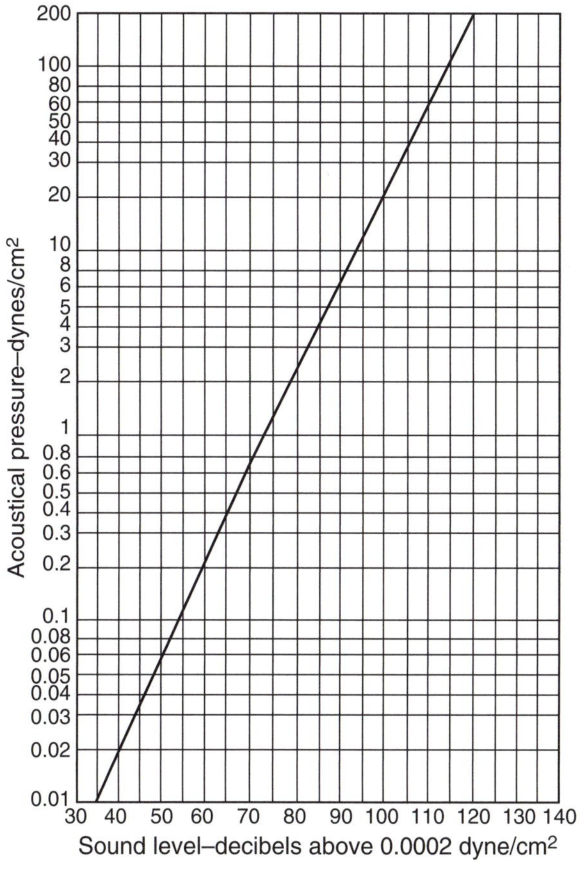

FIGURE 3.6 Sound pressure level versus acoustic pressure.

3.5 SOUND SYSTEM QUANTITIES AND DESIGN FORMULAS

Various quantities used for sound system design are defined as follows:

D_1. D_1 is the distance between the microphone and the loudspeaker, Fig. 3.7.

D_2. D_2 is the distance between the loudspeaker and the farthest listener, Fig. 3.7.

D_o. D_o is the distance between the talker (sound source) and the farthest listener, Fig. 3.7.

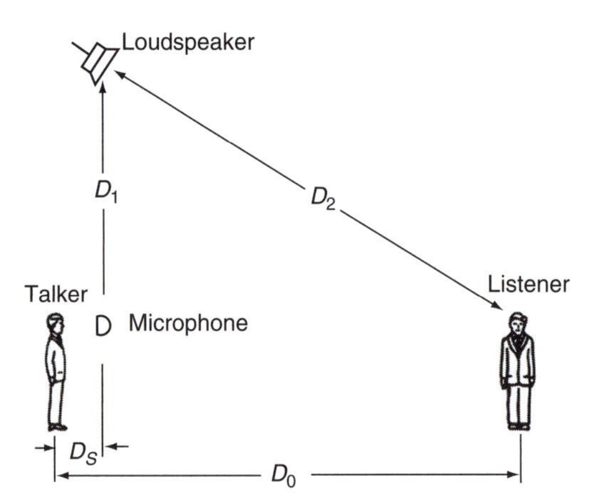

FIGURE 3.7 Definitions of sound system dimensions.

***D_s*.** D_s is the distance between the talker (sound source) and the microphone, Fig. 3.7.

***D_L*.** D_L is the *limiting distance* and is equal to 3.16 D_c for 15%*Alcons* in a room with a reverberation time of 1.6 s. This means that D_2 cannot be any longer than D_L if *Alcons* is to be kept at 15% or less. As the RT_{60} increases or the required %*Alcons* decreases, D_2 becomes less than D_L.

***EAD*.** The *equivalent acoustic distance* (*EAD*) is the maximum distance from the talker that produces adequate loudness of the unamplified voice. Often an *EAD* of 8 ft is used in quiet surroundings as it is the distance at which communications can be understood comfortably. Once the *EAD* has been determined, the sound system is designed to produce that level at every seat in the audience.

***D_c*.** *Critical distance* (D_c) is the point in a room where the direct sound and reverberant sound are equal. D_c is found by the equation:

$$D_c = 0.141\frac{QRM}{N} \tag{3.45}$$

where:

Q is the directivity of the sound source,
R is the room constant,

$a_{1,2...n}$ are the individual absorption coefficients of the areas,
S is the total surface area.

MFP. The *mean-free path* (*MFP*) is the average distance between reflections in a space. *MFP* is found by:

$$MFP = \frac{4V}{S} \tag{3.60}$$

where:
V is the space volume,
S is the space surface area.

ΔD_x. ΔD_x is an arbitrary level change associated with the specific distance from the Hopkins-Stryker equation so that:

$$\Delta D_x = -10\log\left[\frac{Q}{4\pi D_x^2} + \frac{4N}{Sa}\right] \tag{3.61}$$

In semirever berant rooms, Peutz describes ΔD_x as:

$$\Delta D_x = -10\log\left[\frac{Q}{4\pi D_x^2} + \frac{4N}{Sa}\right] + \frac{0.734{**}\sqrt{V}}{h\,\mathrm{RT}_{60}}\log\frac{D_x > D_c}{D_c} \tag{3.62}$$

**200 for SI units
where:
h is the ceiling height.

SNR. *SNR* is the acoustical *signal-to-noise ratio*. The signal-to-noise ratio required for intelligibility is:

$$SNR = 35\left(\frac{2 - \log \%Alcons}{2 - \log 9RT_{60}}\right) \tag{3.63}$$

SPL. *SPL* is the *sound pressure level* in dB-SPL re 0.00002 N/m^2. *SPL* is also called L_p.

Max Program Level. *Max program level* is the maximum program level attainable at a specific point from the available input power. Max program level is:

$$program\ level_{max} = 10\log\frac{watts_{avail}}{10} - (\Delta D_2 - \Delta D_{ref}) + L_{sens} \tag{3.64}$$

L_{sens}. *Loudspeaker sensitivity* (L_{sens}) is the on-axis *SPL* output of the loudspeaker with a specified power input and at a specified distance. The most common L_{sens} are at 4 ft, 1 W and 1 m, and 1 W.

Sa. *Sa* is the *total absorption* in sabines of all the surface areas times their absorption.

dB-SPL$_T$. The *dB-SPL$_T$* is the talker's or sound source's *sound pressure level.*

dB-SPL$_D$. The *dB-SPL$_D$* is the desired *sound pressure level.*

dB-SPL. The *dB-SPL* is the *sound pressure level* in decibels.

EIN. *EIN* is the *equivalent input noise.*

$$EIN = -198\,\text{dB} + 10\log BW + 10\log Z - 6\,\text{dB} - 20\log 0.775 \tag{3.65}$$

where:

BW is the bandwidth,

Z is the impedance.

Thermal Noise. *Thermal noise* is the noise produced in any resistance, including standard resistors. Any resistance that is at a temperature above absolute zero generates noise due to the thermal agitation of free electrons in the material. The magnitude of the noise can be calculated from the resistance, absolute temperature, and equivalent noise bandwidth of the measuring system. A completely

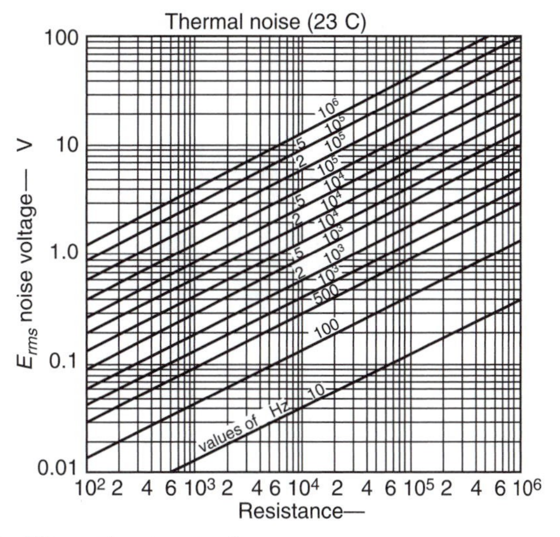

FIGURE 3.8 Thermal noise graph.

noise-free amplifier whose input is connected to its equivalent source resistance will have noise in its output equal to the product of amplification and source resistor noise. This noise is said to be the *theoretical minimum.*

Fig. 3.8 provides a quick means for determining the rms value of thermal noise voltage in terms of resistance and circuit bandwidth.

For practical calculations, especially those in which the resistive component is constant across the band-width of interest, use:

$$E_{rms} = \sqrt{4 \times 10^{-23}(T)(f_1 - f_2)R} \tag{3.66}$$

where:

$f_1 - f_2$ is the 3 dB bandwidth,

R is the resistive component of the impedance across which the noise is developed,

T is the absolute temperature in K.

***RT*₆₀.** RT_{60} is the time required for an interrupted steady-state signal in a space to decay 60 dB. RT_{60} is normally calculated using one of the following equations: the classic Sabine method, the Norris-Eyring modification of the Sabine equation, and the Fitzroy equation. The Fitzroy equation is best used when the walls in the *X*, *Y*, and *Z* planes have very different absorption materials on them.

Sabine:

$$RT_{60} = 0.049 ** \frac{V}{S\overline{a}} \tag{3.67}$$

** 0.161 for SI units.

Norris-Eyring:

$$RT_{60} = 0.049 ** \frac{V}{-S \ln(1 - \overline{a})} \tag{3.68}$$

** 0.161 for SI units.

Fitzroy:

$$RT_{60} = \frac{0.049 ** V}{S^2} \left[\frac{2XY}{-\ln(1 - \overline{a}_{XY})} + \frac{2XZ}{-\ln(1 - \overline{a}_{XZ})} + \frac{2YZ}{-\ln(1 - \overline{a}_{YZ})}\right] \tag{3.69}$$

** 0.161 for SI units.

where:

V is the room volume,
S is the surface area,
a is the total absorption coefficient,
X is the space length,
Y is the space width,
Z is the space height.

Signal Delay. *Signal delay* is the time required for a signal, traveling at the speed of sound, to travel from the source to a specified point in space:

$$SD = \frac{Distance}{c} \tag{3.70}$$

where:

SD is the signal delay in milliseconds,
c is the speed of sound.

3.6 ISO NUMBERS

"Preferred Numbers were developed in France by Charles Renard in 1879 because of a need for a rational basis for grading cotton rope. The sizing system that resulted from his work was based upon a geometric series of mass per unit length such that every fifth step of the series increased the size of rope by a factor of ten." (From the American National Standards for Preferred Numbers). This same system of preferred numbers is used today in acoustics. The one-twelfth, one-sixth, one-third, one-half, two-thirds, and one octave preferred center frequency numbers are not the exact n series number. The exact n series number is found by the equation:

$$n \text{ Series number } = 10^{\frac{1}{n}}\left(10^{\frac{1}{n}}\right)\left(10^{\frac{1}{n}}\right)\ldots \tag{3.71}$$

where:

n is the ordinal numbers in the series.

For instance, the third n number for a 40 series would be

$$10^{\frac{1}{40}}\left(10^{\frac{1}{40}}\right)\left(10^{\frac{1}{40}}\right) = 1.1885022$$

The preferred ISO number is 1.18. Table 3.4 is a table of preferred International Standards Organization (ISO) numbers.

TABLE 3.4 Internationally Preferred ISO Numbers

1/12 oct.	1/6 oct.	1/3 oct.	1/2 oct.	2/3 oct.	1/1 oct.	
40 ser.	**20 ser.**	**10 ser.**	**6⅔ ser.**	**5 ser.**	**3⅓ ser**	**Exact Value**
1.00	1.00	1.00	1.00	1.00	1.00	1.000000000
1.06						1.059253725
1.12	1.12					1.122018454
1.18						1.188502227
1.25	1.25	1.25				1.258925411
1.32						1.333521431

TABLE 3.4 Internationally Preferred ISO Numbers—Contd.

1/12 oct.	1/6 oct.	1/3 oct.	1/2 oct.	2/3 oct.	1/1 oct.	
40 ser.	**20 ser.**	**10 ser.**	**6⅓ ser.**	**5 ser.**	**3⅓ ser**	**Exact Value**
1.40	1.40					1.412537543
1.50						1.496235654
1.60	1.60	1.60		1.60		1.584893190
1.70						1.678804015
1.80	1.80					1.778279406
1.90						1.883649085
2.00	2.00	2.00	2.00		2.00	1.995262310
2.12						2.113489034
2.24	2.24					2.238721132
2.36						2.371373698
2.50	2.50	2.50		2.50		2.511886423
2.65						2.660725050
2.80	2.80		2.80			2.818382920
3.00						2.985382606
3.15	3.15	3.15				3.162277646
3.35						3.349654376
3.55	3.55					3.548133875
3.75						3.758374024
4.00	4.00	4.00	4.00	4.00	4.00	3.981071685
4.25						4.216965012
4.50	4.50					4.466835897
4.75						4.731512563
5.00	5.00	5.00				5.011872307
5.30						5.308844410
5.60	5.60		5.60			5.623413217
6.00						5.956621397
6.30	6.30	6.30		6.30		6.309573403

TABLE 3.4 Internationally Preferred ISO Numbers—Contd.

1/12 oct.	1/6 oct.	1/3 oct.	1/2 oct.	2/3 oct.	1/1 oct.	
40 ser.	**20 ser.**	**10 ser.**	**6⅔ ser.**	**5 ser.**	**3⅓ ser**	**Exact Value**
6.70						6.683439130
7.10	7.10					7.079457794
7.50						7.498942039
8.00	8.00	8.00	8.00		8.00	7.943282288
8.50						8.413951352
9.00	9.00					8.912509312
9.50						9.440608688

3.7 GREEK ALPHABET

The *Greek alphabet* plays a major role in the language of engineering and sound. Table 3.5 shows the Greek alphabet and the terms that are commonly symbolized by it.

3.8 AUDIO STANDARDS

Audio standards are defined by the Audio Engineering Society (AES), Table 3.6, and the International Electrotechnical Commission (IEC), Table 3.7.

3.9 AUDIO FREQUENCY RANGE

The audio spectrum is usually considered the frequency range between 20 Hz and 20 kHz, Fig. 3.9. In reality, the upper limit of hearing pure tones is between 12 kHz and 18 kHz, depending on the person's age and sex and how well the ears have been trained and protected against loud sounds. Frequencies above 20 kHz cannot be heard as a sound, but the effect created by such frequencies (i.e., rapid rise time) can be heard.

The lower end of the spectrum is more often felt than heard as a pure tone. Frequencies below 20 Hz are difficult to

TABLE 3.5 Greek Alphabet

Name	Upper Case		Lower Case	
alpha	A		α	absorption factor, angles, angular acceleration, attenuation constant, common-base current amplification factor, deviation of state parameter, temperature coefficient of linear expansion, temperature coefficient of resistance, thermal expansion coefficient, thermal diffusivity
beta	B		β	angles, common-emitter current amplification factor, flux density, phase constant, wavelength constant
gamma	Γ		γ	electrical conductivity, Grueneisen parameter
delta	Δ	decrement increment	δ	angles, damping coefficient (decay constant), decrement, increment, secondary-emission ratio
epsilon	E	electric field intensity	ε	capacitivity, dielectric coefficient, electron energy, emissivity, permittivity, base of natural logarithms (2.71828)
zeta	Z		ζ	Chemical potential, dielectric susceptibility (intrinsic capacitance), efficiency, hysteresis, intrinsic impedance of a medium, intrinsic standoff ratio
eta	H		η	
theta	Θ	angles, thermal resistance	θ	angle of rotation, angles, angular phase displacement, reluctance, transit angle
iota	I		ι	
kappa	K	coupling coefficient	κ	susceptibility
lambda	Λ		λ	line density of charge, permeance, photosensitivity, wavelength
mu	M		μ	amplification factor, magnetic permeability, micron, mobility, permeability, prefix micro

TABLE 3.5 Greek Alphabet—Contd.

Name	Upper Case		Lower Case	
nu	N		ν	Reluctivity
xi	Ξ		ξ	
omicron	O		ο	
pi	Π		π	Peltier coefficient, ratio of circumference to diameter (3.1416)
rho	P		ρ	reflection coefficient, reflection factor, resistivity, volume density of electric charge
sigma	Σ	summation	σ	conductivity, Stefan-Boltzmann constant, surface density of charge
tau	T	period	τ	propagation constant, Thomson coefficient, time constant, time-phase displacement, transmission factor
upsilon	Y	admittance	Υ	
phi	Φ	magnetic flux, radiant flux	ϕ	angles, coefficient of performance, contact potential, magnetic flux, phase angle, phase displacement
chi	X			angles
Psi	Ψ	angles	ψ	dielectric flux, displacement flux, phase difference
omega	Ω	resistance	ω	angular frequency, angular velocity, solid angle

TABLE 3.6 AES Standards

Standards and Recommended Practices, Issued Jan 2007	
AES2-1984: (r2003)	AES recommended practice—Specification of loudspeaker components used in professional audio and sound reinforcement
AES3-2003:	AES recommended practice for digital audio engineering—Serial transmission format for two-channel linearly represented digital audio data (Revision of AES3-1992, including subsequent amendments)

TABLE 3.6 AES Standards—Contd.

Standards and Recommended Practices, Issued Jan 2007	
AES5-2003:	AES recommended practice for professional digital audio—Preferred sampling frequencies for applications employing pulse-code modulation (revision of AES5-1997)
AES6-1982: (r2003)	Method for measurement of weighted peak flutter of sound recording and reproducing equipment
AES7-2000: (r2005)	AES standard for the preservation and restoration of audio recording—Method of measuring recorded fluxivity of magnetic sound records at medium wavelengths (Revision of AES7-1982)
AES10-2003:	AES recommended practice for digital audio engineering—Serial Multichannel Audio Digital Interface (MADI) (Revision of AES10-1991)
AES11-2003:	AES recommended practice for digital audio engineering—Synchronization of digital audio equipment in studio operations. (Revision of AES11-1997)
AES14-1992: (r2004)	AES standard for professional audio equipment—Application of connectors, part 1, XLR-type polarity and gender
AES15-1991: (w2002)	AES recommended practice for sound-reinforcement systems—Communications interface (PA-422) (Withdrawn: 2002)
AES17-1998: (r2004)	AES standard method for digital audio engineering—Measurement of digital audio equipment (Revision of AES17-1991)
AES18-1996: (r2002)	AES recommended practice for digital audio engineering—Format for the user data channel of the AES digital audio interface. (Revision of AES18-1992)
AES19-1992: (w2003)	AES-ALMA standard test method for audio engineering—Measurement of the lowest resonance frequency of loudspeaker cones (With-drawn: 2003)
AES20-1996: (r2002)	AES recommended practice for professional audio—Subjective evaluation of loudspeakers
AES22-1997: (r2003)	AES recommended practice for audio preservation and restoration—Storage and handling—Storage of polyester-base magnetic tape

TABLE 3.6 AES Standards—Contd.

Standards and Recommended Practices, Issued Jan 2007	
AES24-1-1999: (w2004)	AES standard for sound system control—Application protocol for controlling and monitoring audio devices via digital data networks—Part 1: Principles, formats, and basic procedures (Revision of AES24-1-1995)
AES24-2-tu: (w2004)	PROPOSED DRAFT AES standard for sound system control—Application protocol for controlling and monitoring audio devices via digital data networks—Part 2, data types, constants, and class structure (for Trial Use)
AES26-2001:	AES recommended practice for professional audio—Conservation of the polarity of audio signals (Revision of AES26-1995)
AES27-1996: (r2002)	AES recommended practice for forensic purposes—Managing recorded audio materials intended for examination
AES28-1997: (r2003)	AES standard for audio preservation and restoration—Method for estimating life expectancy of compact discs (CD-ROM), based on effects of temperature and relative humidity (includes Amendment 1-2001)
AES31-1-2001: (r2006)	AES standard for network and file transfer of audio—Audio-file transfer and exchange Part 1: Disk format
AES31-2-2006:	AES standard on network and file transfer of audio—Audio-file transfer and exchange—File format for transferring digital audio data between systems of different type and manufacture
AES31-3-1999:	AES standard for network and file transfer of audio—Audio-file transfer and exchange—Part 3: Simple project interchange
AES32-tu:	PROPOSED DRAFT AES standard for professional audio interconnections—Fibre optic connectors, cables, and characteristics (for Trial Use)
AES33-1999: (w2004)	AES standard—For audio interconnections—Database of multiple—program connection configurations (Withdrawn: 2004)

TABLE 3.6 AES Standards—Contd.

Standards and Recommended Practices, Issued Jan 2007	
AES35-2000: (r2005)	AES standard for audio preservation and restoration—Method for estimating life expectancy of magneto-optical (M-O) disks, based on effects of temperature and relative humidity
AES38-2000: (r2005)	AES standard for audio preservation and restoration—Life expectancy of information stored in recordable compact disc systems—Method for estimating, based on effects of temperature and relative humidity
AES41-2000: (r2005)	AES standard for digital audio—Recoding data set for audio bit-rate reduction
AES42-2006:	AES standard for acoustics—Digital interface for microphones
AES43-2000: (r2005)	AES standard for forensic purposes—Criteria for the authentication of analog audio tape recordings
AES45-2001:	AES standard for single program connectors—Connectors for loudspeaker-level patch panels
AES46-2002:	AES standard for network and file transfer of audio Audio-file transfer and exchange, Radio traffic audio delivery extension to the broadcast-WAVE-file format
AES47-2006:	AES standard for digital audio—Digital input-output interfacing—Transmission of digital audio over asynchronous transfer mode (ATM) networks
AES48-2005:	AES standard on interconnections—Grounding and EMC practices—Shields of connectors in audio equipment containing active circuitry
AES49-2005:	AES standard for audio preservation and restoration—Magnetic tape—Care and handling practices for extended usage
AES50-2005:	AES standard for digital audio engineering—High-resolution multichannel audio inter-connection
AES51-2006:	AES standard for digital audio—Digital input-output interfacing—Transmission of ATM cells over Ethernet physical layer
AES52-2006:	AES standard for digital audio engineering—Insertion of unique identifiers into the AES3 transport stream

TABLE 3.6 AES Standards—Contd.

Standards and Recommended Practices, Issued Jan 2007	
AES53-2006:	AES standard for digital audio—Digital input-output interfacing—Sample-accurate timing in AES47
AES-1id-1991: (r2003)	AES information document—Plane wave tubes: design and practice
AES-2id-1996: (r2001)	AES information document for digital audio engineering—Guidelines for the use of the AES3 interface
AES-3id-2001: (r2006)	AES information document for digital audio engineering—Transmission of AES3 formatted data by unbalanced coaxial cable (Revision of AES-3id-1995)
AES-4id-2001:	AES information document for room acoustics and sound reinforcement systems—Characterization and measurement of surface scattering uniformity
AES-5id-1997: (r2003)	AES information document for room acoustics and sound reinforcement systems—Loudspeaker modeling and measurement—Frequency and angular resolution for measuring, presenting, and predicting loudspeaker polar data
AES-6id-2000:	AES information document for digital audio—Personal computer audio quality measurements
AES-10id-2005:	AES information document for digital audio engineering—Engineering guidelines for the multichannel audio digital interface, AES10 (MADI)
AES-R1-1997:	AES project report for professional audio—Specifications for audio on high-capacity media
AES-R2-2004:	AES project report for articles on professional audio and for equipment specifications—Notations for expressing levels (Revision of AES-R2-1998)
AES-R3-2001:	AES standards project report on single program connector—Compatibility for patch panels of tip-ring-sleeve connectors
AES-R4-2002:	AES standards project report. Guidelines for AES Recommended practice for digital audio engineering—Transmission of digital audio over asynchronous transfer mode (ATM) networks

TABLE 3.6 AES Standards—Contd.

Standards and Recommended Practices, Issued Jan 2007	
AES-R6-2005:	AES project report—Guidelines for AES standard for digital audio engineering—High-resolution multi-channel audio interconnection (HRMAI)
AES-R7-2006:	AES standards project report—Considerations for accurate peak metering of digital audio signals

TABLE 3.7 IEC Standards

IEC Number	IEC Title
IEC 60038	IEC standard voltages
IEC 60063	Preferred number series for resistors and capacitors
IEC 60094	Magnetic tape sound recording and reproducing systems
IEC 60094-5	Electrical magnetic tape properties
IEC 60094-6	Reel-to-reel systems
IEC 60094-7	Cassette for commercial tape records and domestic use
IEC 60096	Radio-frequency cables
IEC 60098	Rumble measurement on vinyl disc turntables
IEC 60134	Absolute maximum and design ratings of tube and semi-conductor devices
IEC 60169	Radio-frequency connectors
IEC 60169-2	Unmatched coaxial connector (Belling-Lee TV Aerial Plug)
IEC 60169-8	BNC connector, 50 ohm
IEC 60169-9	SMC connector, 50 ohm
IEC 60169-10	SMB connector, 50 ohm
IEC 60169-15	N connector, 50 ohm or 75 ohm
IEC 60169-16	SMA connector, 50 ohm
IEC 60169-16	TNC connector, 50 ohm
IEC 60169-24	F connector, 75 ohm
IEC 60179	Sound level meters
IEC 60228	Conductors of insulated cables

TABLE 3.7 IEC Standards—Contd.

IEC Number	IEC Title
IEC 60268	Sound system equipment
IEC 60268-1	General
IEC 60268-2	Explanation of general terms and calculation methods
IEC 60268-3	Amplifiers
IEC 60268-4	Microphones
IEC 60268-5	Loudspeakers
IEC 60268-6	Auxiliary passive elements
IEC 60268-7	Headphones and earphones
IEC 60268-8	Automatic gain control devices
IEC 60268-9	Artificial reverberation, time delay, and frequency shift equipment
IEC 60268-10	Peak program level meters
IEC 60268-11	Application of connectors for the interconnection of sound system components
IEC 60268-12	Application of connectors for broadcast and similar use
IEC 60268-13	Listening tests on loudspeakers
IEC 60268-14	Circular and elliptical loudspeakers; outer frame diameters and mounting dimensions
IEC 60268-16	Objective rating of speech intelligibility by speech transmission index
IEC 60268-17	Standard volume indicators
IEC 60268-18	Peak program level meters—Digital audio peak level meter
IEC 60297	19-inch rack
IEC 60386	Wow and flutter measurement (audio)
IEC 60417	Graphical symbols for use on equipment
IEC 60446	Wiring colors
IEC 60574	Audio-visual, video, and television equipment and systems
IEC 60651	Sound level meters
IEC 60908	Compact disk digital audio system
IEC 61043	Sound intensity meters with pairs of microphones

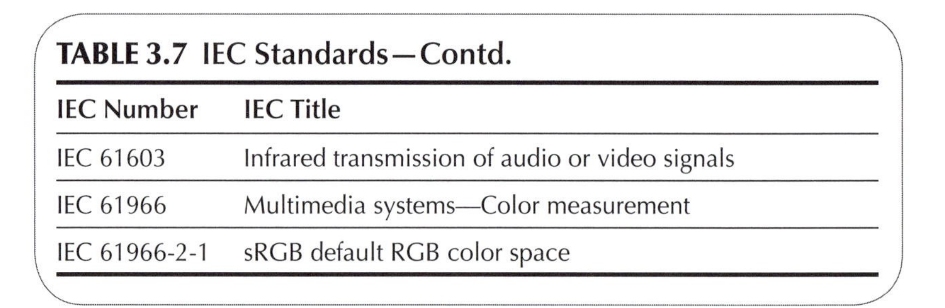

TABLE 3.7 IEC Standards—Contd.

IEC Number	IEC Title
IEC 61603	Infrared transmission of audio or video signals
IEC 61966	Multimedia systems—Color measurement
IEC 61966-2-1	sRGB default RGB color space

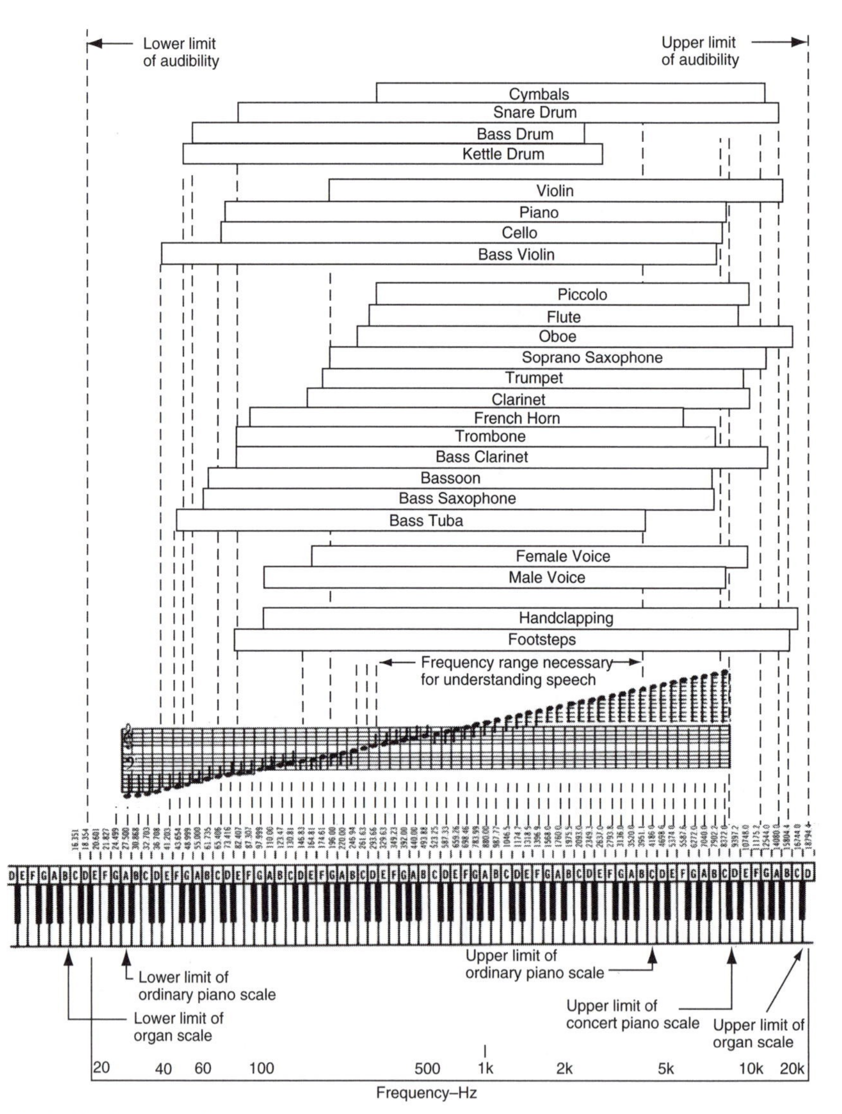

FIGURE 3.9 Audible frequency range.

reproduce. Often the reproducer actually reproduces the second harmonic of the frequency, and the brain translates it back to the fundamental.

3.10 COMMON CONVERSION FACTORS

Conversion from U.S. to SI units can be made by multiplying the U.S. unit by the conversion factors in Table 3.8. To convert from SI units to U.S. units, divide by the conversion factor.

TABLE 3.8 U.S. to SI Units Conversion Factors

U.S. Unit	Multiplier	SI Unit
Length		
ft	3.048000×10^{-1}	m
mi	1.609344×10^{3}	m
in	2.540000×10^{-2}	m
Area		
ft^2	9.290304×10^{-2}	m^2
in^2	6.451600×10^{-4}	m^2
yd^2	8.361274×10^{-1}	m^2
Capacity/volume		
in^3	1.638706×10^{-5}	m^3
ft^3	2.831685×10^{-2}	m^3
liquid gal	3.785412×10^{-3}	m^3
Volume/mass		
ft^3/lb	6.242796×10^{-2}	m^3/kg
in^3/lb	3.612728×10^{-5}	m^3/kg
Velocity		
ft/h	4.466667×10^{-5}	m/s
in/s	2.540000×10^{-2}	m/s
mi/h	4.470400×10^{-1}	m/s

TABLE 3.8 U.S. to SI Units Conversion Factors—Contd.

U.S. Unit	Multiplier	SI Unit
Mass		
oz	2.834952×10^{-2}	kg
lb	4.535924×10^{-1}	kg
Short Ton (2000 lb)	9.071847×10^{2}	kg
Long Ton (2240 lb)	1.016047×10^{3}	kg
Mass/volume		
oz/in^3	1.729994×10^{3}	kg/m^3
lb/ft^3	1.601846×10^{1}	kg/m^3
lb/in^3	2.767990×10^{4}	kg/m^3
lb/U.S. Gal	1.198264×10^{2}	kg/m^3
Acceleration		
ft/s^2	3.048000×10^{-1}	m/s^2
Angular Momentum		
lb f^2/s	4.214011×10^{-2}	$kg.m^2/s$
Electricity		
A•h	3.600000×10^{3}	C
Gs	1.000000×10^{-4}	T
Mx	1.000000×10^{-8}	Wb
Mho	1.000000×10^{0}	S
Oe	7.957747×10^{1}	A/m
Energy (Work)		
Btu	1.055056×10^{3}	J
eV	1.602190×10^{-19}	J
W•h	3.600000×10^{3}	J
erg	1.000000×10^{-7}	J
Cal	4.186800×10^{0}	J

TABLE 3.8 U.S. to SI Units Conversion Factors—Contd.

U.S. Unit	Multiplier	SI Unit
Force		
dyn	1.000000×10^{-5}	N
lbf	4.448222×10^{0}	N
pdl	1.382550×10^{-1}	N
Heat		
Btu/ft^2	1.135653×10^{4}	J/m^2
Btu/lb	2.326000×10^{3}	J/hg
Btu/(h•ft^2•°F) or k (thermal conductivity)	1.730735×10^{0}	W/m•K
Btu/(h•f^2•°F) or C (thermal conductance)	5.678263×10^{0}	W/m^2•K
Btu/(lb °F) or c (heat capacity)	4.186800×10^{3}	J/kg•K
°F•h•ft^2/Btu or R (thermal resistance)	1.761102×10^{-1}	K•m^2/W
cal	4.186000×10^{0}	J
cal/g	4.186000×10^{3}	J/kg
Light		
cd (candle power)	1.000000×10^{0}	cd (candela)
fc	1.076391×10^{1}	lx
fL	3.426259×10^{0}	cd/m^2
Moment of Inertia		
lb.ft^2	4.214011×10^{-2}	kg•m^2
Momentum		
lb.ft/s	1.382550×10^{-1}	kg•m/s
Power		
Btu/h	2.930711×10^{-1}	W
erg/s	1.000000×10^{-7}	W

TABLE 3.8 U.S. to SI Units Conversion Factors—Contd.

U.S. Unit	Multiplier	SI Unit
Power (Contd.)		
hp (550 ft/lb/s)	7.456999×10^{2}	W
hp (electric)	7.460000×10^{2}	W
Pressure		
atm (normal atmosphere)	1.031250×10^{5}	Pa
bar	1.000000×10^{5}	Pa
in Hg@ 60°F	3.376850×10^{3}	Pa
dyn/cm^2	1.000000×10^{-1}	Pa
cm Hg@ 0°C	1.333220×10^{3}	Pa
lbf/f^2	4.788026×10^{1}	Pa
pdl/ft^2	1.488164×10^{0}	Pa
Viscosity		
cP	1.000000×10^{-3}	Pa.s
lb/ft.s	1.488164×10^{0}	Pa.s
ft^2/s	9.290304×10^{-2}	m^2/s
Temperature		
°C	$t_C + 273.15$	K
°F	$(t_F + 459.67)/1.8$	K
°R	$t_R/1.8$	K
°F	$(t_F - 32)/1.8$ or $(t_F - 32) \times (5/9)$	°C
°C	$1.8 (t_C) + 32$ or $(t_C \times (9/5)) + 32$	°F

3.11 TECHNICAL ABBREVIATIONS

Many units or terms in engineering have abbreviations accepted either by the U.S. government or by acousticians, audio consultants, and engineers. Table 3.9 is a list of many of these abbreviations. Symbols for multiple and submultiple prefixes are shown in Table 3.1.

TABLE 3.9 Recommended Abbreviations

Unit or Term	Symbol or Abbreviation
1000 electron volts	keV
A-weighted sound-pressure level in decibels	dBA
absorption coefficient	a
ac current	Iac
ac volt	Vac
acoustic intensity	I_a
Acoustical Society of America	ASA
adaptive delta pulse code modulation	ADPCM
admittance	Y
advanced access control system	AACS
advanced audio coding	AAC
advanced encryption standard	AES
advanced technology attachment	ATA
Advanced Television Systems Committee	ATSC
alien crosstalk margin computation	ACMC
alien far-end crosstalk	AFEXT
alien near-end crosstalk	ANEXT
all-pass filter	APF
alternating current	ac
aluminum steel polyethylene	ASP
ambient noise level	ANL
American Broadcasting Company	ABC
American Federation of Television and Radio Artists	AFTRA
American National Standards Institute	ANSI
American Society for Testing and Materials	ASTM
American Society of Heating, Refrigeration and Air Conditioning Engineers	ASHRAE
American Standard Code for Information Interchange	ASCII

TABLE 3.9 Recommended Abbreviations—Contd.

Unit or Term	Symbol or Abbreviation
American Standards Association	ASA
American wire gauge	AWG
Americans with Disabilities Act	ADA
ampere	A
ampere-hour	Ah
ampere-turn	At
amplification factor	μ
amplitude modulation	AM
analog to digital	A/D
analog-to-digital converter	ADC
angstrom	Å
antilogarithm	antilog
apple file protocol	AFP
appliance wiring material	AWM
articulation index	AI
assisted resonance	AR
assistive listening devices	ALD
assistive listening systems	ALS
asymmetric digital subscriber line	ADSL
asynchronous transfer mode	ATM
atmosphere normal atmosphere technical atmosphere	atm at
atomic mass unit (unified)	u
attenuation to crosstalk ratio	ACR
Audio Engineering Society	AES
audio erase	AE
audio frequency	AF
audio high density	AHD
audio over IP	AoIP

TABLE 3.9 Recommended Abbreviations—Contd.

Unit or Term	Symbol or Abbreviation
audio/video receivers	AVR
automated test equipment	ATE
automatic frequency control	AFC
automatic gain control	AGC
automatic level control	ALC
automatic volume control	AVC
auxiliary	aux
available bit rate	ABR
available input power	AIP
avalanche photodiodes	APD
average	avg
average absorption coefficient	a
average amplitude	A_{avg}
average power	P_{avg}
backlight compensation	BLC
backward-wave oscillator	BWO
balanced current amplifier	BCA
balanced to unbalanced (Bal-Un)	Balun
bandpass filter	BPF
bandpass in hertz	f_{BP}
bandwidth	BW
bar	bar
barn	b
basic rate interface ISDN	BRI
baud	Bd
Bayonet Neill-Concelman	BNC
beat-frequency oscillator	BFO
bel	B
binary coded decimal	BCD

TABLE 3.9 Recommended Abbreviations—Contd.

Unit or Term	Symbol or Abbreviation
binary phase shift keying	BPSK
binaural impulse response	BIR
bipolar junction transistor	BJT
bit	b
bit error rate	BER
bits per second	bps
blue minus luminance	B-Y
breakdown voltage	BV
British Standards Institution	BSI
British thermal unit	Btu
building automation systems	BAS
bulletin board service	BBS
butadiene-acrylonitrile copolymer rubber	NBR
calorie (International Table calorie)	cal_{IT}
calorie (thermochemical calorie)	cal_{th}
Canadian Electrical Code	CEC
Canadian Standards Association	CSA
candela	cd
candela per square foot	cd/ft^2
candela per square meter	cd/m^2
candle	cd
capacitance; capacitor	C
capacitive reactance	X_C
carrier-sense multiple access with collision detection	CSMA/CD
carrier less amplitude phase modulation	CAP
cathode-ray oscilloscope	CRO
cathode-ray tube	CRT
cd universal device format	CD-UDF

TABLE 3.9 Recommended Abbreviations—Contd.

Unit or Term	Symbol or Abbreviation
centimeter	cm
centimeter-gram-second	CGS
central office	CO
central processing unit	CPU
certified technology specialist	CTS
charge coupled device	CCD
charge transfer device	CTD
chlorinated polyethylene	CPE
circular mil	cmil
citizens band	CB
closed circuit television	CCTV
coated aluminum polyethylene basic sheath	Alpeth
coated aluminum, coated steel	CASPIC
coated aluminum, coated steel, polyethylene	CACSP
coercive force	H_c
Columbia Broadcasting Company	CBS
Comité Consultatif International des Radio-communications	CCIR
commercial online service	COLS
Commission Internationale de l'Eclairage	CIE
common mode rejection or common mode rejection ratio	CMR, CMRR
Communications Cable and Connectivity Cable Association	CCCA
compact disc	CD
compact disc digital audio	CD-DA
compact disc interactive	CD-I
compression/decompression algorithm	CODEC
computer aided design	CAD
conductor flat cable	FCFC

TABLE 3.9 Recommended Abbreviations—Contd.

Unit or Term	Symbol or Abbreviation
consolidation point	CP
constant angular velocity	CAV
constant bandwidth	CB
constant bandwidth filter	CBF
constant bit rate	CBR
constant linear velocity	CLV
constant percentage bandwidth	CPB
constant-amplitude phase-shift	CAPS
Consumer Electronics Association	CEA
contact resistance stability	CRS
content scrambling system	CSS
Continental Automated Building Association	CABA
continuous wave	CW
coulomb	C
coverage angle	C
critical bands	CB
critical distance	D_c
Cross Interleave Reed Solomon Code	CIRC
cross linked polyethylene	XLPE
cubic centimeter	cm^3
cubic foot	ft^3
cubic foot per minute	ft^3/min
cubic foot per second	ft^3/s
cubic inch	in^3
cubic meter	m^3
cubic meter per second	m^3/s
cubic yard	yd^3
curie	Ci
Custom Electronics Design and Installation Association	CEDIA

TABLE 3.9 Recommended Abbreviations—Contd.

Unit or Term	Symbol or Abbreviation
customer service representative	CSR
cycle per second	Hz
cyclic redundancy check	CRC
data encryption standard	DES
data over cable service interface	DOCSIS
dc current	I_{dc}
dc voltage	V_{dc}
decibel	dB
decibel ref to one milliwatt	dBm
decibels with a reference of 1 V	dBV
deferred procedure calls	DPC
degree (plane angle)	...°
degree Celsius	°C
degree Fahrenheit	°F
denial of service	DoS
dense wave division multiplexing	DWDM
depth of discharge	DOD
descriptive video service	DVS
Deutsche Industrie Normenausschuss	DIN
Deutsches Institute fur Normung	DIN
device under test	DUT
diameter	diam
dielectric absorption	DA
differential thermocouple voltmeter	DTVM
digital audio broadcasting	DAB
digital audio stationary head	DASH
digital audio tape	DAT

TABLE 3.9 Recommended Abbreviations—Contd.

Unit or Term	Symbol or Abbreviation
Digital Audio Video Council	DAVIC
digital audio workstations	DAW
digital compact cassette	DCC
digital data storage	DDS
digital home standard	DHS
digital light processing	DLP
digital micromirror device	DMD
digital phantom power	DPP
digital rights management	DRM
digital room correction	DRC
digital satellite system	DSS
digital signal processing	DSP
Digital Subscriber Line	DSL
digital sum value	DSV
digital to analog	D/A
digital TV	DTV
digital versatile disc	DVD
digital VHS	D-VHS
digital video	DV
digital video broadcasting	DVB
digital visual interface	DVI
digital voltmeter	DVM
digital-to-analog converter	DAC
direct broadcast	DBS
direct broadcast satellite	DBS
direct current	dc
direct current volts	V_{dc}
direct memory access	DMA

TABLE 3.9 Recommended Abbreviations—Contd.

Unit or Term	Symbol or Abbreviation
direct metal mastering	DMM
direct satellite broadcast	DSB
direct sound level in dB	dB_{DIR}
direct sound pressure level	L_D
direct stream digital	DSD
direct stream transfer	DST
direct time lock	DTL
direct to disk mastering	DDM
direct to home	DTH
directivity factor	Q
directivity index	DI
Discrete Fourier Transform	DFT
discrete multitone	DMT
display data channel	DDC
display power management signaling	DPMS
dissipation factor	DF
double sideband	DSB
dual expanded plastic insulated conductor	DEPIC
dual in-line package	DIP
dual-tone multi-frequency	DTMF
dynamic host configuration protocol	DHCP
dynamic host control protocol	DHCP
dynamic noise reduction	DNR
dyne	dyn
EIA microphone sensitivity rating	G_M
eight-to-fourteen modulation	EFM
electrical metallic tubing	EMT
electrical power required	EPR
electrocardiograph	EKG

TABLE 3.9 Recommended Abbreviations—Contd.

Unit or Term	Symbol or Abbreviation
electromagnetic compatibility	ECM
electromagnetic interference	EMI
electromagnetic radiation	emr
electromagnetic unit	EMU
electromechanical relay	EMR
electromotive force	emf
electron volt	eV
electronic data processing	EDP
Electronic Field Production	EFP
Electronic Industries Alliance	EIA
Electronic Industries Association (obsolete)	EIA
electronic iris	E.I.
electronic music distribution	EMD
electronic news gathering	ENG
electronic voltohmmeter	EVOM
electronvolt	eV
electrostatic unit	ESU
Emergency Broadcast System	EBS
end of life vehicle	ELV
energy density level	L_W
energy frequency curve	EFC
energy level	L_E
energy-time-curve	ETC
Enhanced Definition Television	EDTV
enhanced direct time lock	DTLe
enhanced IDE	EIDE
environmental protection agency	EPA
equal level far end crosstalk	ELFEXT
equalizer	EQ

TABLE 3.9 Recommended Abbreviations—Contd.

Unit or Term	Symbol or Abbreviation
equipment distribution area	EDA
equivalent acoustic distance	EAD
equivalent input noise	EIN
equivalent rectangular bandwidth	ERB
equivalent resistance	R_{eq}
equivalent series inductance	ESL
equivalent series resistance	ESR
error checking and correcting random-access memory	ECC RAM
ethylene-propylene copolymer rubber	EPR
ethylene-propylene-diene monomer rubber	EPDM
European Broadcasting Union	EBU
expanded polyethylene-polyvinyl chloride	XPE-PVC
extended data out RAM EDO	RAM
extra-high voltage	EHV
extremely high frequency	EHF
extremely low frequency	ELF
far end crosstalk	FEXT
farad	F
Fast Discrete Fourier Transform	FDFT
Fast Fourier Transform	FFT
fast link pulses	FLPs
Federal Communications Commission	FCC
feedback stability margin	FSM
fiber data distributed interface	FDDI
fiber distribution frame	FDF
fiber optic connector	FOC
fiber optics	FO
fiber to the curb	FTTC

TABLE 3.9 Recommended Abbreviations—Contd.

Unit or Term	Symbol or Abbreviation
fiber to the home	FTTH
field programmable gate array	FPGA
field-effect transistor	FET
file transfer protocal	FTP
finite difference	FD
finite difference time domain	FDTD
finite impulse response	FIR
fire alarm and signal cable	FAS
flame retardant ethylene propylene	FREP
flame retarded thermoplastic elastomer	FR-TPE
flexible OLED	FOLED
flexible organic light emitting diode	FOLED
fluorinated ethylene propylene	FEP
flux density	B
foot	ft/ [']
foot per minute	ft/min
foot per second	ft/s
foot per second squared	ft/s^2
foot pound-force	ft.lbf
foot poundal	ft.dl
footcandle	fc
footlambert	fL
forward error correction	FEC
four-pole, double-throw	4PDT
four-pole, single-throw	4PST
fractional part of	FRC
frame check sequence	FCS
frequency modulation	FM
frequency time curve	FTC

TABLE 3.9 Recommended Abbreviations—Contd.

Unit or Term	Symbol or Abbreviation
frequency-shift keying	FSK
frequency; force	f
frequently asked question	FAQ
full scale	FS
function indicator panel	FIP
gallon	gal
gallon per minute	gal/min
gauss	G
General Services Administration	GSA
gigacycle per second	GHz
gigaelectronvolt	GeV
gigahertz	GHz
gilbert	Gb
gram	g
Greenwich Mean Time	GMT
ground	GND
gypsum wallboard	GWB
head related transfer function	HRTF
Hearing Loss Association of America	HLAA
heating, ventilating, and air conditioning	HVAC
henry	H
hertz	Hz
high bit-rate digital subscriber line	HDSL
high definition—serial digital interface	HD-SDI
high definition multimedia interface	HDMI
high frequency	HF
high voltage	HV
high-bandwidth digital content protection	HDCP

TABLE 3.9 Recommended Abbreviations—Contd.

Unit or Term	Symbol or Abbreviation
high-definition multimedia interface	HDMI
high-definition television	HDTV
high-density linear converter system	HDLCS
high-pass filter	HPF
high-speed cable data service	HSCDS
high-speed parallel network technology	HIPPI
Home Automation and Networking Association	HANA
horizontal connection point	HCP
horizontal distribution areas	HDAs
horsepower	hp
hour	h
hybrid fiber/coaxial	HFC
hypertext markup language	HTML
hypertext transfer protocol	HTTP
ignition radiation suppression	IRS
impedance (magnitude)	*Z*
impulse response	IR
in the ear	ITE
inch	in, ["]
inch per second	in/s
independent consultants in audiovisual technology	ICAT
independent sideband	ISB
index matching gel	IMG
index of refraction	IOR
inductance	L
inductance-capacitance	LC
inductive reactance	X_L
inductor	L

TABLE 3.9 Recommended Abbreviations—Contd.

Unit or Term	Symbol or Abbreviation
infinite impulse response	IIR
infrared	IR
initial signal delay	ISD
initial time delay gap	ITDG
inner hair cells	IHC
input-output	I/O
inside diameter	ID
Institute of Electrical and Electronic Engineers	IEEE
Institute of Radio Engineers	IRE
instructional television fixed service	ITFS
Insulated Cable Engineers Association	ICEA
insulated gate field effect transistor	IGFET
insulated gate transistor	IGT
insulation displacement connector	IDC
insulation resistance	IR
integrated circuit	IC
integrated detectors/preamplifiers	IDP
integrated device electronics	IDE
integrated electronic component	IEC
integrated network management system	INMS
Integrated Services Digital Network	ISDN
intelligent power management system	IPM™
intensity level	L_I
interaural cross-correlation	IACC
interaural cross-correlation coefficient	IACC
interaural intensity difference	IID
interaural level difference	ILD
interaural phase difference	IPD

TABLE 3.9 Recommended Abbreviations—Contd.

Unit or Term	Symbol or Abbreviation
interaural time difference	ITD
intermediate frequency	IF
intermodulation	IM
intermodulation distortion	IM or IMD
international building code	IBC
International Communication Industries Association	ICIA
International Electrotechnical Commission	IEC
International Electrotechnical Engineers	IEEE
International Organization for Standardization	ISO
International Radio Consultative Committee	CCIR
International Standard Recording Code	ISRC
International Standards Organization	ISO
International Telecommunication Union	ITU
Internet Engineering Task Force	IETF
internet group management protocol	IGMP
internet protocol	IP
internet service provider	ISP
interrupted feedback (foldback)	IFB
inverse discrete Fourier transform	IDFT
IP Over Cable Data Network	IPCDN
ISDN digital subscriber line	IDSL
Japanese Standards Association	JSA
Joint Photographic Experts Group	JPEG
joule	J
joule per kelvin	J/K
junction field effect transistor	JFET
just noticeable difference	JND

TABLE 3.9 Recommended Abbreviations—Contd.

Unit or Term	Symbol or Abbreviation
kelvin	K
kilocycle per second	kHz
kiloelectronvolt	keV
kilogauss	kG
kilogram	kg
kilogram-force	kgf
kilohertz	kHz
kilohm	kΩ
kilojoule	kJ
kilometer	km
kilometer per hour	km/h
kilovar	kvar
kilovolt (1000 volts)	kV
kilovolt-ampere	kVA
kilowatt	kW
kilowatthour	kWh
knot	kn
lambert	L
large area systems	LAS
large-scale hybrid integration	LSHI
large-scale integration	LSI
lateral efficiency	LE
lateral fraction	LF
leadership in energy and environmental design	LEED
least significant bit	LSB
left, center, right, surrounds	LCRS
light amplification by stimulated emission of radiation	LASER

TABLE 3.9 Recommended Abbreviations—Contd.

Unit or Term	Symbol or Abbreviation
light dependent resistor	LDR
light emitting diode	LED
linear time invariant	LTI
liquid crystal display	LCD
liquid crystal on silicon	LCoS
listening environment diagnostic recording	LEDR
liter	l
liter per second	l/s
live end—dead end	LEDE
local area network	LAN
local exchange carrier	LEC
local multipoint distribution service	LMDS
logarithm	log
logarithm, natural	ln
long play	LP
look up table	LUT
loudspeaker sensitivity	L_{sens}
low frequency	LF
low frequency effects	LFE
low power radio services	LPRS
low-frequency effects	LFE
low-pass filter	LPF
lower sideband	LSB
lumen	lm
lumen per square foot	lm/ft^2
lumen per square meter	lm/m^2
lumen per watt	lm/W
lumen second	lm.s

TABLE 3.9 Recommended Abbreviations—Contd.

Unit or Term	Symbol or Abbreviation
lux	lx
magneto hydrodynamics	MHD
magneto-optical recording	MOR
magneto-optics	MO
magnetomotive force	MMF
mail transfer protocol	MTP
main cross connect	MC
main distribution areas	MDA
manufacturing automation protocol	MAP
Mass Media Bureau	MMB
master antenna television	MATV
matched resistance	R_M
maxwell	Mx
mean free path	MFP
media access control	MAC
medium area systems	MAS
medium frequency	MF
megabits per second	Mbps
megabyte	MB
megacycle per second	MHz
megaelectronvolt	MeV
megahertz	MHz
megavolt	MV
megawatt	MW
megohm	MΩ
metal-oxide semiconductor	MOS
metal-oxide semiconductor field-effect transistor	MOSFET
metal-oxide varistor	MOV

TABLE 3.9 Recommended Abbreviations—Contd.

Unit or Term	Symbol or Abbreviation
meter	m
meter-kilogram-second	MKS
metropolitan area network	MAN
microampere	μA
microbar	μbar
microelectromechanical systems	MEMS
microfarad	μF
microgram	μg
microhenry	μH
micrometer	μm
micromho	μmho
microphone	mic
microsecond	μs
microsiemens	μS
microvolt	μV
microwatt	μW
midi time code	MTC
mile (statute)	mi
mile per hour	mi/h
milli	m
milliampere	mA
millibar	mbar
millibarn	mb
milligal	mGal
milligram	mg
millihenry	mH
milliliter	ml
millimeter	mm
millimeter of mercury, conventional	mmHg

TABLE 3.9 Recommended Abbreviations—Contd.

Unit or Term	Symbol or Abbreviation
millisecond	ms
millisiemens	mS
millivolt	mV
milliwatt	mW
minidisc	MD
minute (plane angle)	...'
minute (time)	min
modified rhyme test	MRT
modulation reduction factor	m(F)
modulation transfer function	MTF
modulation transmission function	MTF
mole	mol
most significant bit	MSB
motion drive amplifier	MDA
motion JPEG	M-JPEG
Motion Picture Experts Group	MPEG
moves, adds, and changes	MACs
moving coil	MC
multichannel audio digital interface	MADI
multichannel reverberation	MCR
multichannel audio digital interface	MADI
Multimedia Cable Network System Partners Ltd	MCNS
multiple system operator	MSO
multiple-in/multiple-out	MIMO
multiplier/accumulator	MAC
multipoint distribution system	MDS
multistage noise shaping	MASH
multiuser telecommunications outlet assembly	MUTOA
music cd plus graphics	CD-G

TABLE 3.9 Recommended Abbreviations—Contd.

Unit or Term	Symbol or Abbreviation
musical instrument digital interface	MIDI
mutual inductance	M
nanoampere	nA
nanofarad	nF
nanometer	nm
nanosecond	ns
nanowatt	nW
National Association of Broadcasters	NAB
National Association of the Deaf	NAD
National Broadcasting Company	NBC
National Bureau of Standards	NBS
National Electrical Code	NEC
National Electrical Contractors Association	NECA
National Electrical Manufacturers Association	NEMA
National Fire Protection Association	NFPA
National Institute Of Occupational Safety And Health	NIOSH
National Systems Contractors Association	NSCA
National Television Standards Committee	NTSC
National Television System Committee	NTSC
near end cross talk	NEXT
near-instantaneous companding	NICAM
needed acoustic gain	NAG
negative-positive-negative	NPN
neper	Np
network address translation	NAT
network operations center	NOC
neutral density filter	N/D
newton	N

TABLE 3.9 Recommended Abbreviations—Contd.

Unit or Term	Symbol or Abbreviation
newton meter	N·m
newton per square meter	N/m^2
no-epoxy/no-polish	NENP
noise figure; noise frequency	NF
noise reduction coefficient	NRC
noise voltage	E_n
noise-operated automatic level adjuster	NOALA
nonreturn-to-zero inverted	NRZI
normal link pulses	NLPs
number of open microphone	NOM
numerical aperture	NA
oersted	Oe
Office de Radiodiffusion Television Française	ORTF
ohm	Ω
on-screen manager	OSM™
open system interconnect	OSI
open-circuit voltage	E_o
operational transconductance amplifier	OTA
operations support systems	OSS
opposed current interleaved amplifier	OCIA
optical carrier	OC
optical time domain reflectometer	OTDR
optimized common mode rejection	OCMR
optimum power calibration	OPC
optimum source impedance	OSI
optoelectronic integrated circuit	OEIC
organic light emitting diode	OLED
orthogonal frequency division multiplexing	OFDM
ounce (avoirdupois)	oz

TABLE 3.9 Recommended Abbreviations—Contd.

Unit or Term	Symbol or Abbreviation
outer hair cells	OHC
output level in dB	L_{out}
output voltage	E_{OUT}
outside diameter	OD
oxygen-free, high-conductivity copper	OFHC
pan/tilt/zoom	PTZ
parametric room impulse response	PRIR
pascal	Pa
peak amplitude	A_p
peak program meter	PPM
peak reverse voltage	PRV
peak-inverse-voltage	piv
peak-reverse-voltage	prv
peak-to-peak amplitude	A_{p-p}
percentage of articulation loss for consonants	%Alcons
perfluoroalkoxy	PFA
permanent threshold shift	PTS
personal listening systems	PLS
phase alternation line	PAL
phase angle	I
phase frequency curve	PFC
phase locked loop	PLL
phase modulation	PM
phonemically balanced	PB
physical medium dependent	PMD
pickup	PU
picoampere	pA
picofarad	pF
picosecond	ps

TABLE 3.9 Recommended Abbreviations—Contd.

Unit or Term	Symbol or Abbreviation
picowatt	pW
picture in picture	PIP
pinna acoustic response	PAR
pint	pt
plain old telephone service	POTS
plasma	PDP
plastic insulated conductor	PIC
plate current	I_p
plate efficiency	E_{ff}
plate resistance	r_p
plate voltage	E_p
point-to-point protocol	PPP
polarization beam splitter	PBS
polyethylene	PE
polyethylene aluminum steel polyethylene	PASP
polypropylene	PP
polyurethane	PUR
polyvinyl chloride	PVC
polyvinylidene fluoride	PVDF
positive-negative-positive	PNP
positive, intrinsic, negative	PIN
potential acoustic gain	PAG
pound	lb
pound (force) per square inch. Although the use of the abbreviation psi is common, it is not recommended.	lbf/in^2, psi
pound-force	lbf
pound-force foot	lb·fft

TABLE 3.9 Recommended Abbreviations—Contd.

Unit or Term	Symbol or Abbreviation
poundal	pdl
power backoff	PBO
power calibration area	PCA
power factor	PF
power factor correction	PFC
power level	L_W, dB-PWL
power out	P_o
power over ethernet	PoE
power sourcing equipment	PSE
power sum alien equal level far-end crosstalk	PSAELFEXT
power sum alien far-end crosstalk	PSAFEXT
power sum alien near-end crosstalk	PSANEXT
power sum alien NEXT	PSANEXT
powered devices	PD
preamplifier	preamp
precision adaptive subband coding	PASC
precision audio link	PAL
prefade listen	PFL
primary rate interface ISDN	PRI
printed circuit	PC
private branch exchange	PBX
Professional Education and Training Committee	PETC
programmable gate array	PGA
programmable logic device	PLD
programmable read-only memory	PROM
public switched telephone network	PSTN

TABLE 3.9 Recommended Abbreviations—Contd.

Unit or Term	Symbol or Abbreviation
pulse code modulation	PCM
pulse density modulation	PDM
pulse end modulation	PEM
pulse-amplitude modulation	PAM
pulse-duration modulation	PDM
pulse-frequency-modulation	PFM
pulse-position modulation	PPM
pulse-repetition frequency	PRF
pulse-repetition rate	PRR
pulse-time modulation	PTM
pulse-width modulation	PWM
quadratic residue diffuser	QRD
quality factor	Q
quality of service	QoS
quandrature amplitude modulation	QAM
quart	qt
quarter wave plate	QWP
quaternary phase shift keying	QPSK
rad	rd
radian	rad
radio data service	RDS
radio frequency	RF
radio frequency identification	RFID
radio information for motorists	ARI
radio-frequency interference	RFI
rambus DRAM	Rambus, RDRAM
random access memory	RAM
random-noise generator	RNG

TABLE 3.9 Recommended Abbreviations—Contd.

Unit or Term	Symbol or Abbreviation
rapid speech transmission index	RASTI
reactance	X
read-only memory	ROM
real-time analyzer	RTA
real-time transport protocol	RTP
Recording Industry Association of America	RIAA
recording management area	RMA
red, green, blue	RGB
redundant array of independent disks	RAID
reflection-free zone	RFZ
reflections per second	RPS
Regional Data Center	RDC
registered communication distribution designer	RCDD
remote authentication dial-in user service	RADIUS
report on comments	ROC
report on proposal	ROP
request for proposals	RFP
resistance-capacitance	RC
resistance-inductance-capacitance	RLC
resistor	R
resistor-capacitor	RC
resistor-transistor logic	RTL
resource reservation protocol	RSVP
restriction of hazardous substances	RoHS
return loss	RL
reverberant sound level in dB	L_R
reverberation time	RT_{60}
revolution per minute	r/min, rpm

TABLE 3.9 Recommended Abbreviations—Contd.

Unit or Term	Symbol or Abbreviation
revolution per second	r/s, rps
ripple factor	Y
robust service network	RSN
roentgen	R
room constant	Sa
root-mean-square	rms
root-mean-square voltage	Vrms
rotary head digital audio tape	R-DAT
round conductor flat Cable	RCFC
sample and hold	S/H
sample-rate convertor	SRC
satellite news gathering	SNG
Screen Actors Guild	SAG
screened twisted pair	ScTP
second (plane angle)	"
second (time)	s
second audio program	SAP
secure digital music initiative	SDMI
self-monitoring analysis and reporting technology	SMART
sensitivity	sensi
Sequence Electronique Couleur Avec Memoire	SECAM
serial copy management system	SCMS
serial digital interface	SDI
serial digital video	SDV
service station identifier	SSID
shield current induced noise	SCIN
shielded twisted pair(s)	STP

TABLE 3.9 Recommended Abbreviations—Contd.

Unit or Term	Symbol or Abbreviation
short noise	i_{sn}
short wave	SW
siemens	S
signal delay	SD
signal-to-noise ratio	SNR
silicon controlled rectifier	SCR
simple control protocol	SCP
simple network management protocol	SNMP
single in-line package	SIP
single sideband	SSB
single-pair high bit-rate digital subscriber line	S-HDSL
single-pole, double-throw	SPDT
single-pole, single-throw	SPST
small computer system	SCSI
Society of Automotive Engineers	SAE
Society of Motion Picture & Television Engineers	SMPTE
solid state music	SSM
solid state relay	SSR
song position pointer	SPP
sound absorption average	SAA
sound level meter	SLM
sound pressure in dB	dB_{SPL}
sound pressure level	L_p, SPL
sound transmission class	STC
sound, audiovisual, and video integrators	SAVVI
source resistance	R_s
speech transmission index	STI
square foot	ft^2

TABLE 3.9 Recommended Abbreviations—Contd.

Unit or Term	Symbol or Abbreviation
square inch	in^2
square meter	m^2
square yard	yd^2
standard definition – serial digital interface	SD-SDI
standard definition television	SDTV
standing-wave ratio	SWR
static contact resistance	SCR
static RAM	SRAM
steradian	sr
storage area network	SAN
structural return loss	SRL
structured cabling system	SCS
stubs wire gage	SWG
subminiature A connector	SMA
subminiature B connector	SMB
subminiature C connector	SMC
Subsidiary Communications Authorization	SCA
super audio CD	SACD
super audio compact disc	SACD
super video home system Super	VHS
super-luminescent diode	SLD
super-high frequency	SHF
symmetric digital subscriber line	SDSL
synchronous code division multiple access	S-CDMA
synchronous optical network	SONET
table of contents	TOC
telecommunications enclosure	TE
Telecommunications Industry Association	TIA

TABLE 3.9 Recommended Abbreviations—Contd.

Unit or Term	Symbol or Abbreviation
telecommunications room	TR
television	TV
television interference	TVI
temperature differential	ΔT
temporary threshold shift	TTS
tesla	T
tetrafluoroethylene	TFE
thermal noise	TN, i_{tn}
thermocouple; time constant	TC
thin film transistors	TFT
thousand circular mils	kcmil
three-pole, double-throw	3PDT
three-pole, single-throw	3PST
time	T
time delay spectrometry	TDS
time division multiple access	TDMA
time division multiplexing	TDM
time energy frequency	TEF
timebase corrector	TBC
ton	ton
tonne	t
total harmonic distortion	THD
total harmonic distortion plus noise	THD+N
total sound level in dB	L_T
total surface area	S
transient intermodulation distortion	TIM
transistor-transistor logic	TTL
transmission control protocol/internet protocol	TCP/IP

TABLE 3.9 Recommended Abbreviations—Contd.

Unit or Term	Symbol or Abbreviation
transmission loss	TL
transparent OLED	TOLED
transparent organic light Emitting diode	TOLED
transverse electric	TE
transverse electromagnetic	TEM
transverse magnetic	TM
traveling-wave tube	TWT
TV receive only	TVRO
twisted pair-physical medium dependent	TP-PMD
ultrahigh frequency	UHF
ultraviolet	UV
Underwriters Laboratories, Inc.	UL
uniform building code	UBC
unit interval	UI
unit of absorption	Sabin
universal disc format	UDF
Universal Powerline Association	UPA
universal serial bus	USB
universal service order code	USOC
unshielded twisted pair(s)	UTP
upper sideband	USB
user datagram protocol	UDP
user defined protocol	UDP
vacuum-tube voltmeter	VTVM
variable constellation/multitone modulation	VC/MTM
variable speed oscillator	VSO
variable-frequency oscillator	VFO
velocity of propagation	VP

TABLE 3.9 Recommended Abbreviations—Contd.

Unit or Term	Symbol or Abbreviation
vertical-cavity surface-emitting laser	VCSEL
very high bit rate digital subscriber line	VDSL
very high frequency	VHF
very low frequency	VLF
vibratory acceleration level	L_a
vibratory force level	L_F
vibratory velocity level	L_v
Video Electronics Standards Association	VESA
video graphics array	VGA
video home system	VHS
video on demand	VOD
video RAM	VRAM
virtual local area network	VLAN
virtual private networks	VPN
voice over internet protocol	VoIP
voice over wireless fidelity	VoWi-Fi
volt	V
volt-ohm-milliammeter	VOM
voltage (electromotive force)	E
voltage controlled crystal oscillator	VCXO
voltage gain	μ
voltage standing wave ratio	VSWR
voltage-controlled amplifier	VCA
voltage-controlled oscillator	VCO
voltampere	VA
volume indicator	VI
volume unit	VU

TABLE 3.9 Recommended Abbreviations—Contd.

Unit or Term	Symbol or Abbreviation
watt	W
watt per steradian	W/sr
watt per steradian square meter	$W/(sr.m^2)$
watthour	Wh
wavelength	M
wavelength division multiplexing	WDM
weber	Wb
weighted modulation transmission function	WMTF
wide area network	WAN
windows media audio	WMA
wired equivalent privacy	WEP
wireless access points	WAPs
wireless application protocol	WAP
wireless communications service	WCS
wireless fidelity	WiFi
wireless microwave access	WiMax
World Health Organization	WHO
write once	WO
write once read many	WORM
yard	yd
zone distribution area	ZDA

3.12 SURFACE AREA AND VOLUME EQUATIONS

To find the *surface area and volume* of complex areas, the area can often be divided into a series of simpler areas and handled one at a time. Figs. 3.10–3.17 are equations for various and unusual volumes.

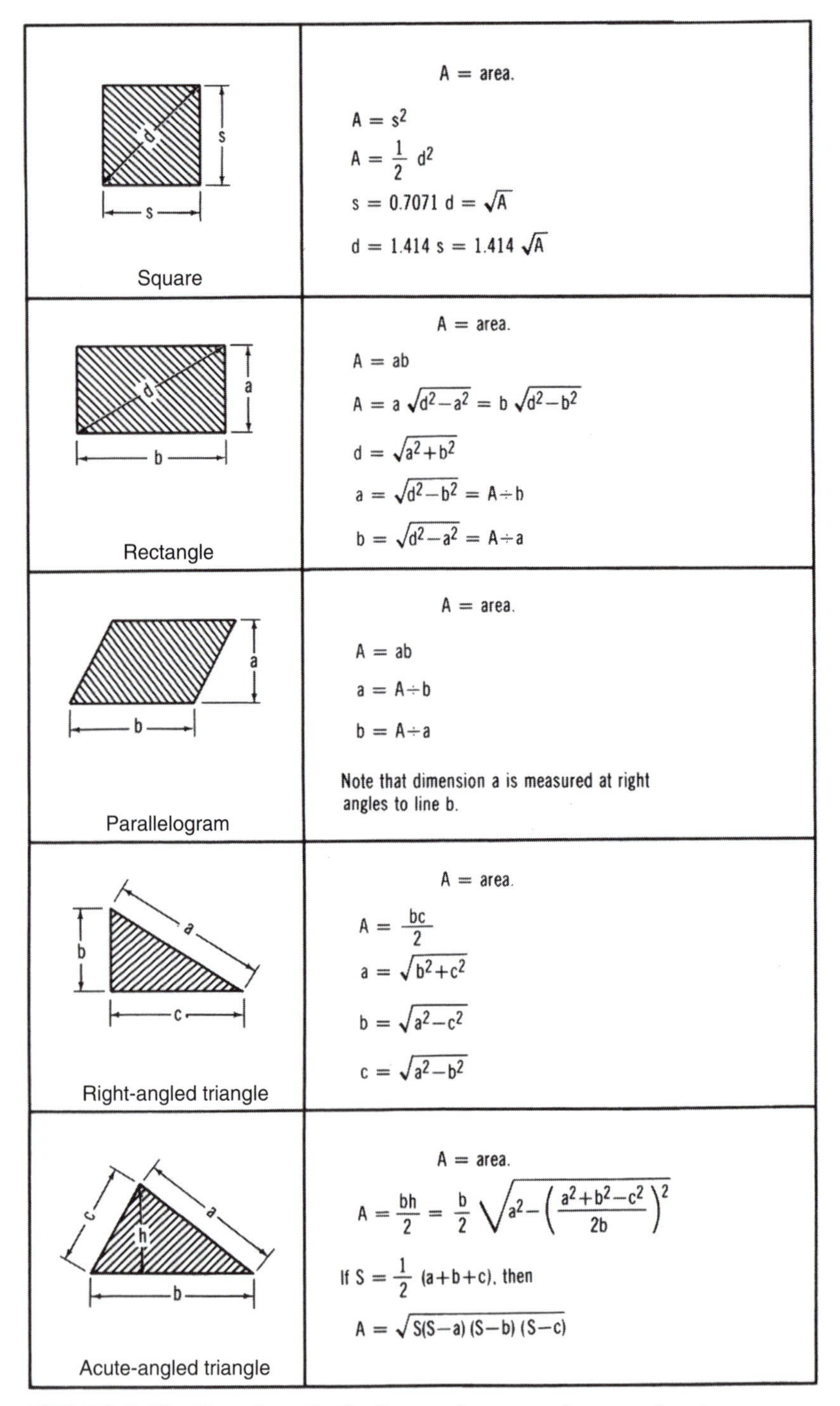

Square	A = area. $A = s^2$ $A = \frac{1}{2} d^2$ $s = 0.7071\, d = \sqrt{A}$ $d = 1.414\, s = 1.414 \sqrt{A}$
Rectangle	A = area. $A = ab$ $A = a\sqrt{d^2-a^2} = b\sqrt{d^2-b^2}$ $d = \sqrt{a^2+b^2}$ $a = \sqrt{d^2-b^2} = A \div b$ $b = \sqrt{d^2-a^2} = A \div a$
Parallelogram	A = area. $A = ab$ $a = A \div b$ $b = A \div a$ Note that dimension a is measured at right angles to line b.
Right-angled triangle	A = area. $A = \frac{bc}{2}$ $a = \sqrt{b^2+c^2}$ $b = \sqrt{a^2-c^2}$ $c = \sqrt{a^2-b^2}$
Acute-angled triangle	A = area. $A = \frac{bh}{2} = \frac{b}{2}\sqrt{a^2-\left(\frac{a^2+b^2-c^2}{2b}\right)^2}$ If $S = \frac{1}{2}(a+b+c)$, then $A = \sqrt{S(S-a)(S-b)(S-c)}$

FIGURE 3.10 Equations for finding surface areas for complex shapes.

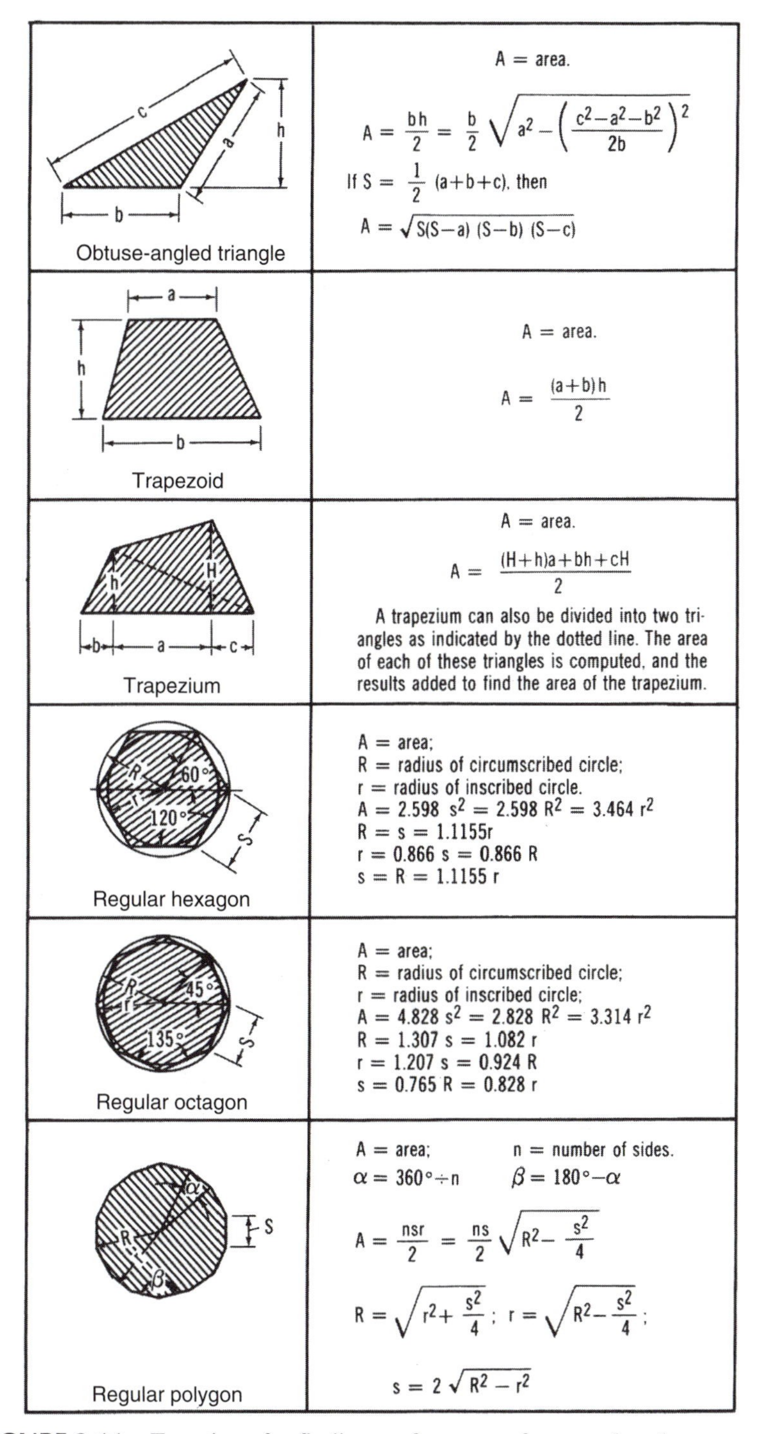

FIGURE 3.11 Equations for finding surface areas for complex shapes.

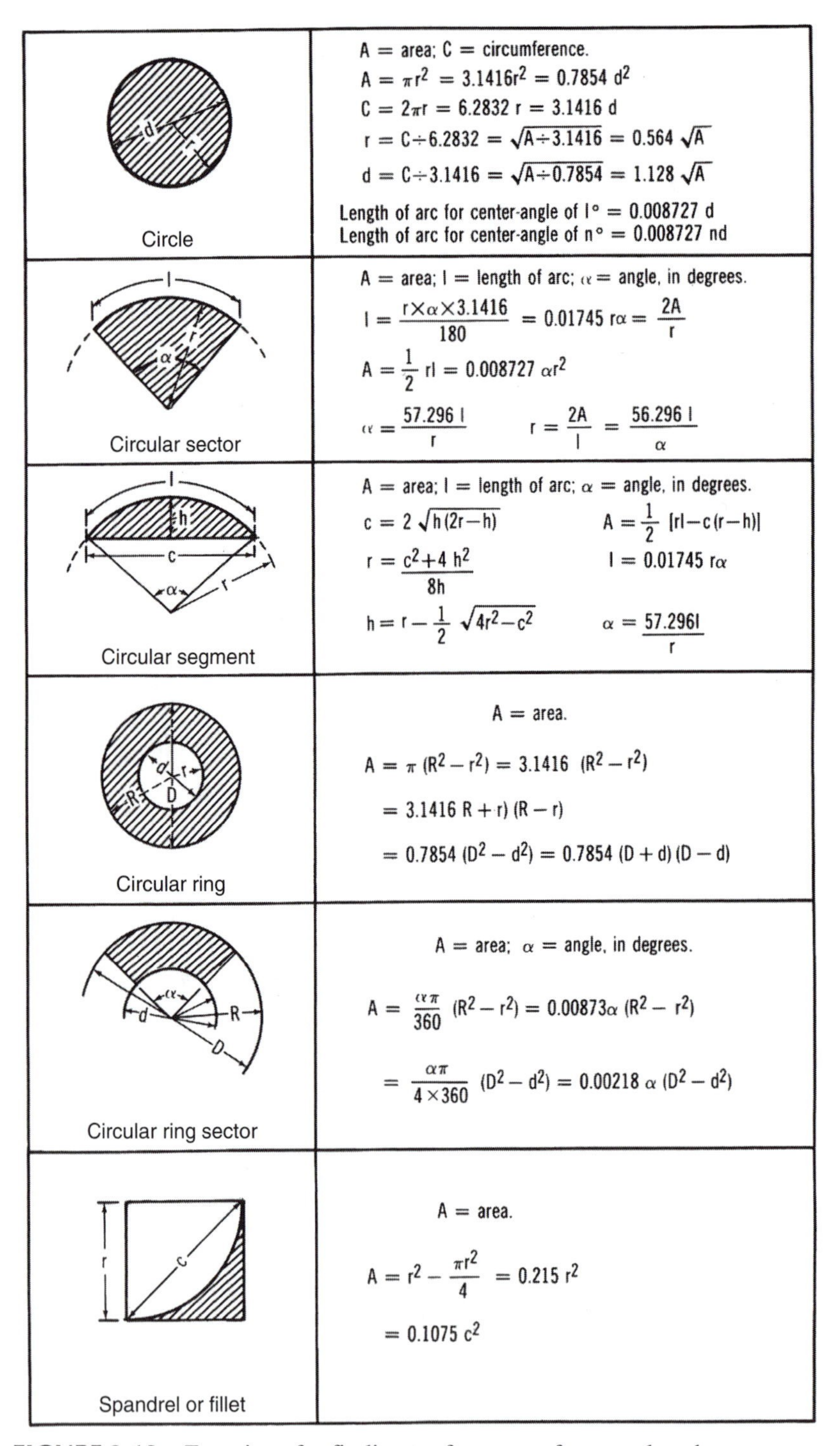

Circle	A = area; C = circumference. $A = \pi r^2 = 3.1416 r^2 = 0.7854\ d^2$ $C = 2\pi r = 6.2832\ r = 3.1416\ d$ $r = C \div 6.2832 = \sqrt{A \div 3.1416} = 0.564\sqrt{A}$ $d = C \div 3.1416 = \sqrt{A \div 0.7854} = 1.128\sqrt{A}$ Length of arc for center-angle of 1° = 0.008727 d Length of arc for center-angle of n° = 0.008727 nd
Circular sector	A = area; l = length of arc; α = angle, in degrees. $l = \frac{r \times \alpha \times 3.1416}{180} = 0.01745\ r\alpha = \frac{2A}{r}$ $A = \frac{1}{2} rl = 0.008727\ \alpha r^2$ $\alpha = \frac{57.296\ l}{r}$ $\quad r = \frac{2A}{l} = \frac{56.296\ l}{\alpha}$
Circular segment	A = area; l = length of arc; α = angle, in degrees. $c = 2\sqrt{h(2r-h)}$ $\quad A = \frac{1}{2}[rl - c(r-h)]$ $r = \frac{c^2 + 4h^2}{8h}$ $\quad l = 0.01745\ r\alpha$ $h = r - \frac{1}{2}\sqrt{4r^2 - c^2}$ $\quad \alpha = \frac{57.2961}{r}$
Circular ring	A = area. $A = \pi(R^2 - r^2) = 3.1416\ (R^2 - r^2)$ $= 3.1416\ R + r)(R - r)$ $= 0.7854\ (D^2 - d^2) = 0.7854\ (D + d)(D - d)$
Circular ring sector	A = area; α = angle, in degrees. $A = \frac{\alpha\pi}{360}(R^2 - r^2) = 0.00873\alpha\ (R^2 - r^2)$ $= \frac{\alpha\pi}{4 \times 360}(D^2 - d^2) = 0.00218\ \alpha\ (D^2 - d^2)$
Spandrel or fillet	A = area. $A = r^2 - \frac{\pi r^2}{4} = 0.215\ r^2$ $= 0.1075\ c^2$

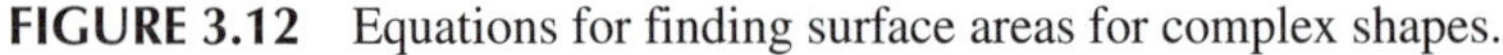

FIGURE 3.12 Equations for finding surface areas for complex shapes.

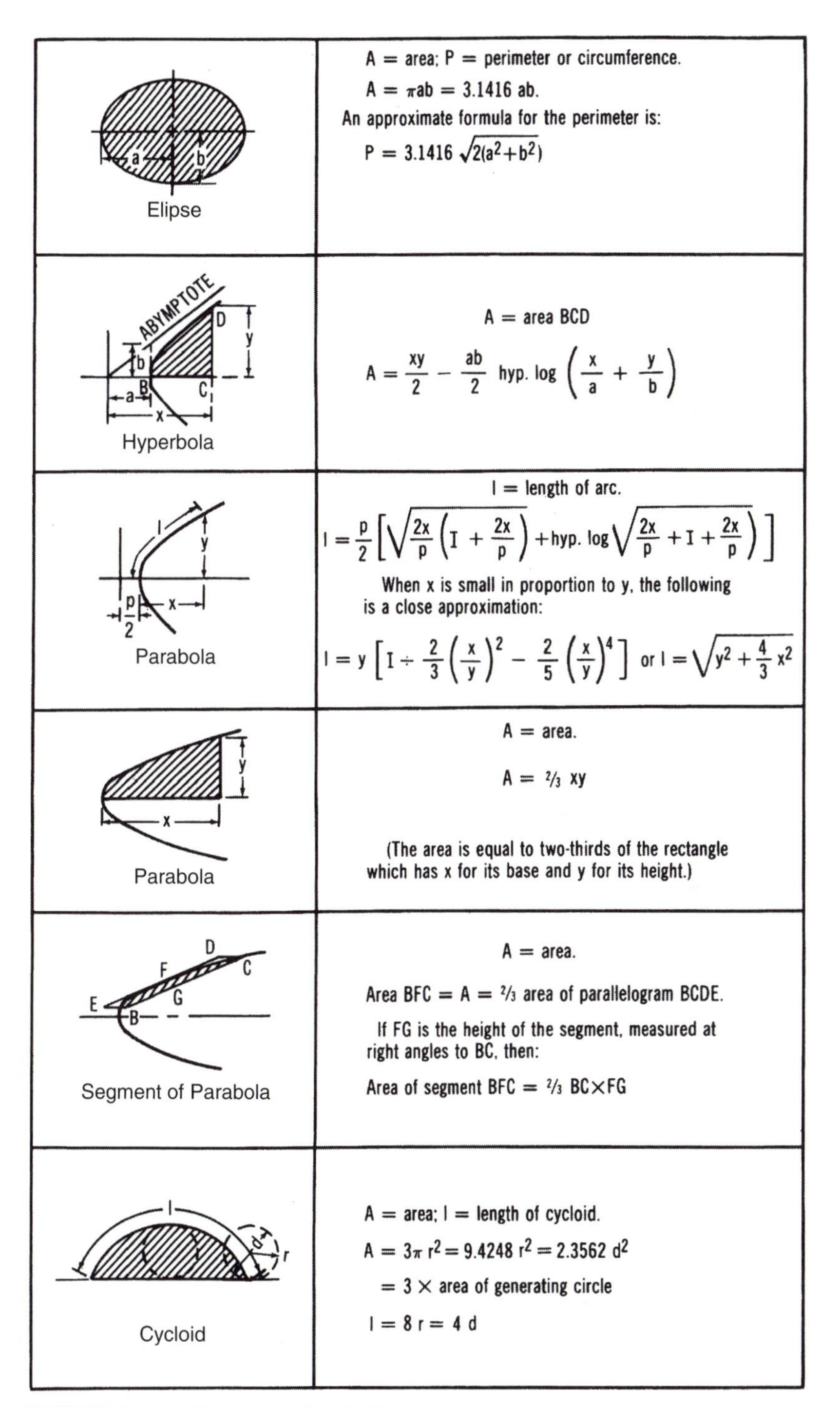

Elipse	A = area; P = perimeter or circumference. $A = \pi ab = 3.1416\ ab$. An approximate formula for the perimeter is: $P = 3.1416\sqrt{2(a^2+b^2)}$
Hyperbola	A = area BCD $A = \frac{xy}{2} - \frac{ab}{2}\ \text{hyp. log}\left(\frac{x}{a} + \frac{y}{b}\right)$
Parabola	l = length of arc. $l = \frac{p}{2}\left[\sqrt{\frac{2x}{p}\left(1 + \frac{2x}{p}\right)} + \text{hyp. log}\left(\sqrt{\frac{2x}{p}} + 1 + \frac{2x}{p}\right)\right]$ When x is small in proportion to y, the following is a close approximation: $l = y\left[1 \div \frac{2}{3}\left(\frac{x}{y}\right)^2 - \frac{2}{5}\left(\frac{x}{y}\right)^4\right]$ or $l = \sqrt{y^2 + \frac{4}{3}x^2}$
Parabola	A = area. $A = \frac{2}{3}xy$ (The area is equal to two-thirds of the rectangle which has x for its base and y for its height.)
Segment of Parabola	A = area. Area BFC = A = ⅔ area of parallelogram BCDE. If FG is the height of the segment, measured at right angles to BC, then: Area of segment BFC = ⅔ BC×FG
Cycloid	A = area; l = length of cycloid. $A = 3\pi r^2 = 9.4248\ r^2 = 2.3562\ d^2$ $= 3 \times$ area of generating circle $l = 8\ r = 4\ d$

FIGURE 3.13 Equations for finding surface areas for complex shapes.

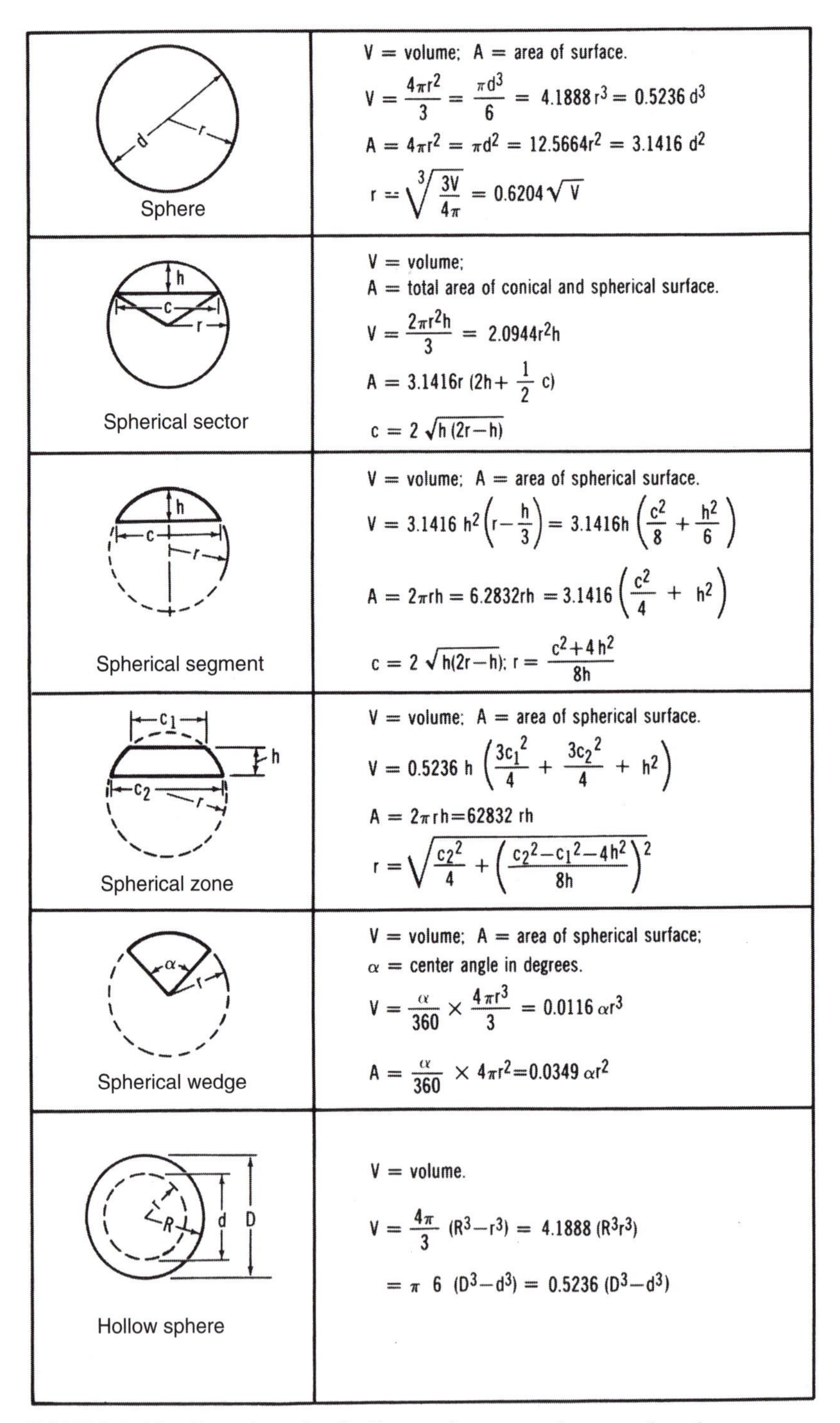

FIGURE 3.14 Equations for finding surface areas for complex shapes.

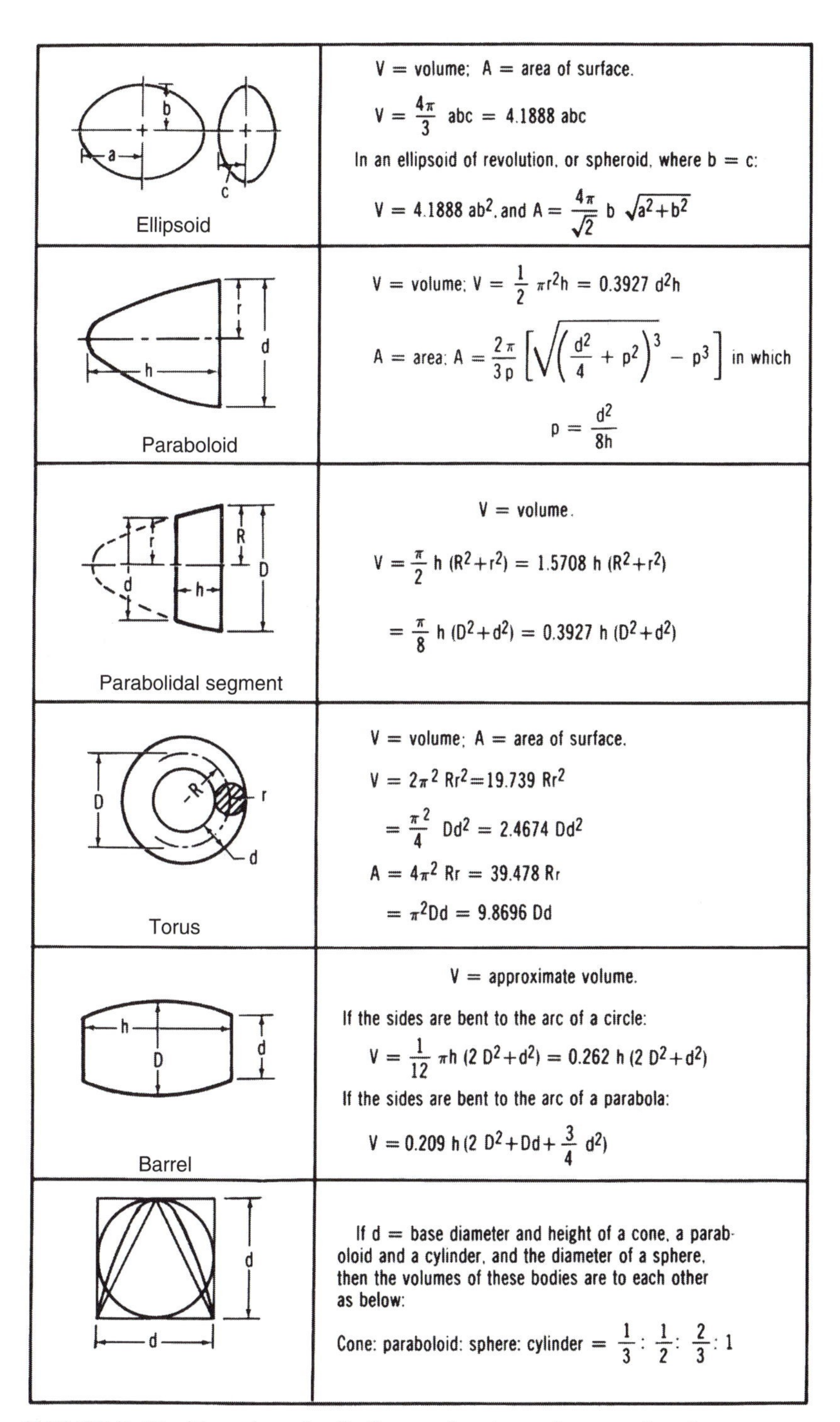

Ellipsoid	V = volume; A = area of surface. $V = \frac{4\pi}{3} abc = 4.1888\, abc$ In an ellipsoid of revolution, or spheroid, where b = c: $V = 4.1888\, ab^2$, and $A = \frac{4\pi}{\sqrt{2}}\, b \sqrt{a^2+b^2}$
Paraboloid	V = volume; $V = \frac{1}{2} \pi r^2 h = 0.3927\, d^2 h$ A = area; $A = \frac{2\pi}{3p}\left[\sqrt{\left(\frac{d^2}{4} + p^2\right)^3} - p^3\right]$ in which $p = \frac{d^2}{8h}$
Parabolidal segment	V = volume. $V = \frac{\pi}{2} h\, (R^2+r^2) = 1.5708\, h\, (R^2+r^2)$ $= \frac{\pi}{8} h\, (D^2+d^2) = 0.3927\, h\, (D^2+d^2)$
Torus	V = volume; A = area of surface. $V = 2\pi^2 Rr^2 = 19.739\, Rr^2$ $= \frac{\pi^2}{4} Dd^2 = 2.4674\, Dd^2$ $A = 4\pi^2 Rr = 39.478\, Rr$ $= \pi^2 Dd = 9.8696\, Dd$
Barrel	V = approximate volume. If the sides are bent to the arc of a circle: $V = \frac{1}{12} \pi h\, (2D^2+d^2) = 0.262\, h\, (2D^2+d^2)$ If the sides are bent to the arc of a parabola: $V = 0.209\, h\, (2D^2+Dd+\frac{3}{4} d^2)$
	If d = base diameter and height of a cone, a paraboloid and a cylinder, and the diameter of a sphere, then the volumes of these bodies are to each other as below: Cone: paraboloid: sphere: cylinder = $\frac{1}{3} : \frac{1}{2} : \frac{2}{3} : 1$

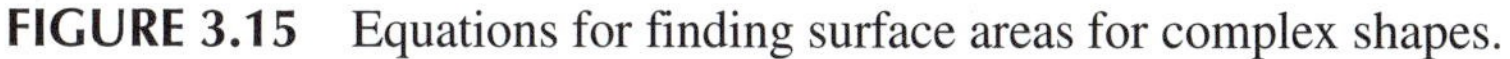

FIGURE 3.15 Equations for finding surface areas for complex shapes.

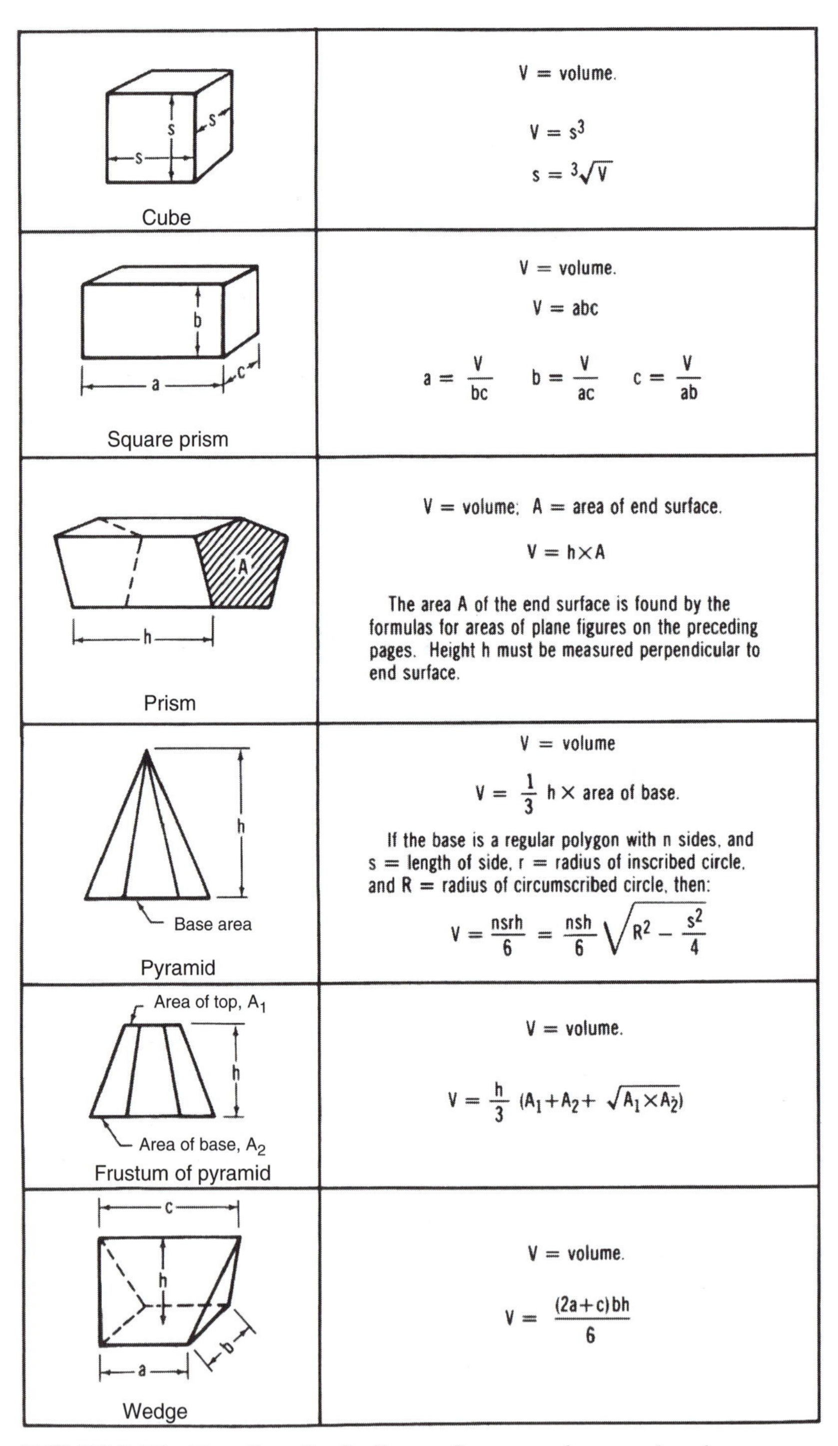

FIGURE 3.16 Equations for finding surface areas for complex shapes.

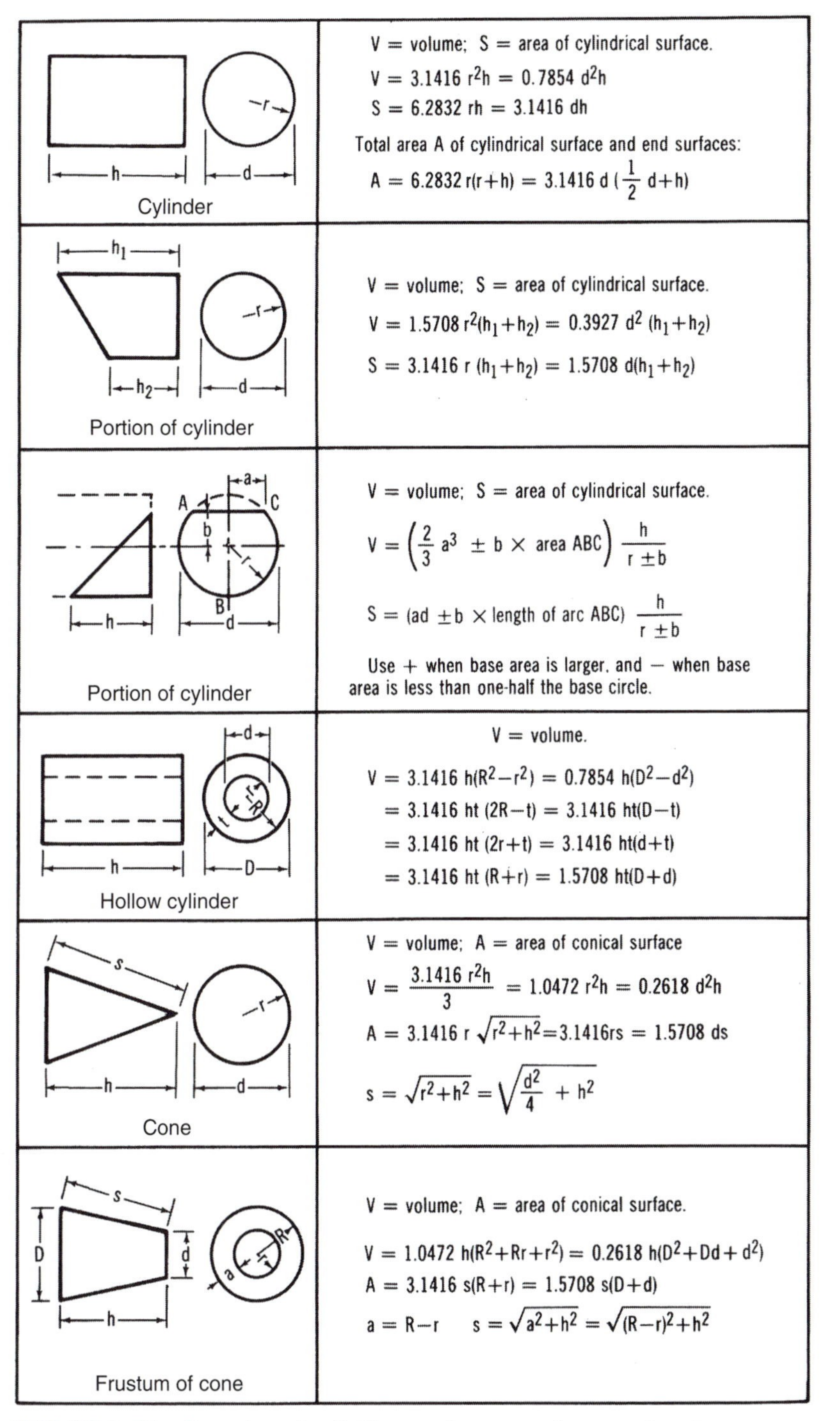

Cylinder	V = volume; S = area of cylindrical surface. $V = 3.1416\, r^2h = 0.7854\, d^2h$ $S = 6.2832\, rh = 3.1416\, dh$ Total area A of cylindrical surface and end surfaces: $A = 6.2832\, r(r+h) = 3.1416\, d\,(\frac{1}{2}\, d+h)$
Portion of cylinder	V = volume; S = area of cylindrical surface. $V = 1.5708\, r^2(h_1+h_2) = 0.3927\, d^2\,(h_1+h_2)$ $S = 3.1416\, r\,(h_1+h_2) = 1.5708\, d(h_1+h_2)$
Portion of cylinder	V = volume; S = area of cylindrical surface. $V = \left(\frac{2}{3}\, a^3 \;\pm b \times \text{area ABC}\right) \frac{h}{r \pm b}$ $S = (ad \;\pm b \times \text{length of arc ABC}) \;\frac{h}{r \pm b}$ Use + when base area is larger, and − when base area is less than one-half the base circle.
Hollow cylinder	V = volume. $V = 3.1416\, h(R^2-r^2) = 0.7854\, h(D^2-d^2)$ $= 3.1416\, ht\,(2R-t) = 3.1416\, ht(D-t)$ $= 3.1416\, ht\,(2r+t) = 3.1416\, ht(d+t)$ $= 3.1416\, ht\,(R+r) = 1.5708\, ht(D+d)$
Cone	V = volume; A = area of conical surface $V = \frac{3.1416\, r^2h}{3} = 1.0472\, r^2h = 0.2618\, d^2h$ $A = 3.1416\, r\,\sqrt{r^2+h^2} = 3.1416rs = 1.5708\, ds$ $s = \sqrt{r^2+h^2} = \sqrt{\frac{d^2}{4} + h^2}$
Frustum of cone	V = volume; A = area of conical surface. $V = 1.0472\, h(R^2+Rr+r^2) = 0.2618\, h(D^2+Dd+d^2)$ $A = 3.1416\, s(R+r) = 1.5708\, s(D+d)$ $a = R-r \qquad s = \sqrt{a^2+h^2} = \sqrt{(R-r)^2+h^2}$

FIGURE 3.17 Equations for finding surface areas for complex shapes.

BIBLIOGRAPHY

Metric Units of Measure and Style Guide, U.S. Metric Association, Inc., Sugarloaf Star Route, Boulder, CO 80302.

V. Antoine, *Guidance for Using the Metric System* (SI Version), Society for Technical Communication, 1010 Vermont Avenue, NW, Washington, DC 20005.

D. Davis, "Synergetic Audio Concepts," vol. 5, no. 2, October 1977.

John L. Feirer, *SI Metric Handbook*, The Metric Company.

Handbook of Electronics Tables and Formulas, 6th Edition, Carmel IN: SAMS, a division of Macmillan Computer Publishing, 1986.

D. Bohn, "The Bewildering Wilderness," *S&VC*, September 2000.

Index

P

R

S